How Science Works: Evolution

Frontispiece

The angler fish is a striking example of adaptation to life in the deep ocean. One of its dorsal spines has evolved into a moveable fleshy lure containing luminescent bacteria. The light attracts prey to within reach of the formidable jaws. The body is capable of expanding enough to swallow prey twice its size. The animal shown is a female angler fish. It is hard for the sexes to meet in the deep ocean so the smaller males bite into a female and stay attached to be fed through her blood vessels.

John Ellis

How Science Works: Evolution

A Student Primer

 Springer

R. John Ellis is an Emeritus Professor of Biological Sciences at the University of Warwick, Coventry UK, CV4 7AL. In 1983 he was elected Fellow of the Royal Society for his research on the biogenesis of chloroplasts. He received an International Gairdner Award in 2004 for fundamental discoveries in chaperone-mediated protein folding, and the Annual Medal of Cell Stress Society International in 2006 for the discovery of the chaperonins. His current interests include neurodegenerative disease caused by protein aggregation, and the effects of macromolecular crowding on cellular processes.

ISBN 978-90-481-3182-2   e-ISBN 978-90-481-3183-9
DOI 10.1007/978-90-481-3183-9
Springer Dordrecht Heidelberg London New York

Library of Congress Control Number: 2009941981

Cover photograph: a land iguana on the Galapagos Islands. These animals share a herbivorous lifestyle with marine iguanas and were described by Charles Darwin during his voyage in HMS *Beagle*. Photograph courtesy of Hugh Woodland.

Printed on acid-free paper

Springer is part of Springer Science+Business Media (www.springer.com)

All our science, measured against reality, is primitive and child-like – and yet it is the most precious thing we have

Albert Einstein

. . . and freedom of thought is best promoted by the gradual illumination of men's minds which follows from the advance of science

Charles Darwin

For Diana and Juliet

Preface

The Importance of Science

When I was growing up, I found myself, like all young people, in a world full of adults telling me stories. Stories about the nature of the world and the place of humans in it, stories about what to believe and how to behave. My problem was that there were many different stories, so how do you decide to choose between them and form your own views? I asked my father about this, and he advised me, when forming my view of the world, not to take any notice of the status of people making particular claims. It did not matter, he said, whether they were called prince or bishop or professor, or whether they wore fancy clothes like mitres or mortar boards. The only thing that mattered, he said, was the quality of the evidence in support of the claims they were making, and that I should be the judge of that quality. By quality, he meant how robust was the evidence? Was it just an opinion derived from what other people said, or was it based solidly on empirical observations anyone could make? In other words, how scientific was the evidence?

This book is based on my father's advice to seek out the best-quality, empirical evidence you can find, and stick to what it implies until better evidence comes along. It is the *quality* of the evidence that you need to learn how to evaluate. In this book I discuss the criteria you should bear in mind when evaluating the evidence for any claim, in the hope that this will help you to develop an independent, critical way of thinking. Reduced to the simplest terms, my advice to you when confronted by people making claims about any subject is not to accept their claims at face value, but to ask "How do you know that?" and rigorously examine the quality of the replies they give.

I decided to write this book after I was asked by Hugh Woodland in 2006 to contribute to a new course in evolution to undergraduates studying biology at the University of Warwick. This request came as a surprise because my research career, before my retirement in 1996, had been in protein biochemistry and I had not taught evolution before. The motivation for devising this new course was partly the feeling that we did not talk about evolution enough at Warwick, given its central importance for all aspects of biology, and partly to combat the resurgence of creationist views that was claimed to be happening in Britain. The latter claim surprised me because evolution had been established as a fact in the 19th century, so I consulted Michael Reiss from the Institute of Education, then on temporary secondment to the Royal Society as its Director of Science Education.

Michael Reiss confirmed to me that increasing numbers of students were entering school and universities who do not accept evolution because their parents do not. He further explained that the response of teachers to such students tended to take one of two courses. They either disparaged and ridiculed the views of these students or they ducked the whole issue, and did not discuss either evolution or creationism. Michael Reiss took the view that both responses are counter-productive, and that a better response to this situation would be to use it as an opportunity to explain how science works and differs from other ways of explaining the world, and hence why creationism is not science.

I began to read about evolution and the issues surrounding it today, and the document that impressed me the most was the online record of the court case held in Dover, York County, Pennsylvania, in 2005. This record can be read at: http://www.pamd.uscourts.gov/kitzmiller/decision.htm. In this case, a group of eleven parents sued their local school Board of Education for requiring that a statement referring to intelligent design as an alternative to evolution must be read to students attending biology lessons, implying that intelligent design is a scientific theory. Intelligent design was defined during this case as meaning that "various forms of life began abruptly through an intelligent agency, with their distinctive features already intact - fish with fins and scales, birds with feathers, beaks and wings etc". The theory of evolution, in contrast, states that all forms of life are related to one another and change with time.

The judge was asked to decide whether intelligent design is a scientific view or a religious view, because the First Amendment of the Constitution of the United States states that "Congress shall make no law respecting an establishment of religion, or prohibiting the free exercise thereof; or abridging the freedom of speech, or of the press". This Amendent thus forbids the teaching of religion in state-funded schools in order to preserve both the freedom of religion and the separation of religion from the state. The judge had to decide how science differs from religion, and to this end several philosophers who had studied this matter gave evidence. The record of the testimony that led the judge to rule that intelligent design is a religious view, and not part of science, is available online, and I found it to be a valuable source of information about the views of philosophers regarding the distinctions between science and religion. References to articles by two of these philosophers, Barbara Forrest and Robert Pennock, are given in the Further Reading list at the end of Chapter 1.

There was a second experience that persuaded me to teach in what, for me, was a new area. I watched an interview by Jeremy Paxman on the BBC TV programme *Newsnight* with the late evolutionary biologist John Maynard Smith. Paxman started the interview by asking Smith "Evolution is just a theory, isn't it?" Smith replied that the evidence for evolution is as good as the evidence for the existence of atoms, but that, on the other hand, he could not rule out the possibility that his whole knowledge of the world had been implanted in his brain ten seconds ago by a capricious supernatural agent. This exchange made me realise that even a person as educated and intelligent as Jeremy Paxman apparently did not realise what the word "theory" means in science or that all scientific knowledge is provisional and subject to change.

My decision to write this book was confirmed by an article that appeared in *The Guardian* newspaper on November 20th, 2008, written by Jim Al-Khalili, a professor of physics and the public engagement in science at the University of Surrey. He wrote

> I do feel strongly that those scientists who have a voice must be doing more than simply popularising their field in order to attract the next generation into science. Yes, this is vital: but it is also vital that we help defend our rational secular society against the rising tide of irrationalism and ignorance. Science communicators, for want of a better term, have a role to play in explaining not just the scientific facts but how science itself works: that it is not just 'another way of explaining the world', and that without it we would still be living in the dark ages.

I have chosen evolution as the example for explaining how science works because it graphically illustrates the issues I wish to address, and because evolution is under increasing attack in some educational establishments. Biologists regard evolution as both a theory and a fact, but evolution is more than just another scientific theory because it challenges those views that suggest humans are basically different from other animals and so can escape the laws of nature. It is this aspect of evolution that makes it so unattractive to many people. But rejecting evolution or any other branch of science means that we reject the best means we have found so far to understand ourselves and our place in the world. Evolution is the greatest story ever told.

You may find some elements of this book controversial, so let me say right at the start that it is not my intention to tell you what to think. My intention is to help you to learn *how* to think, by discussing the sorts of consideration you should bear in mind when formulating your own views about the nature of the world, especially about how scientists study the world. The term "world" in this book means everything that we experience.

Coventry, UK   John Ellis

Acknowledgements

I thank the following for commenting critically on drafts of parts of this book: John Allen, Hugh Cable, Peter Csermely, Diana Ellis, David Epstein, Robert Freedman, Walter Gratzer, David Hodgson, Richard Johnson, Ken Joy, Harry Kroto, Robert Old, Geoff Oxford, Kevin Padian, Michael Reiss, Robert Spooner, Janet Thornton, Peter Waister, and Lewis Wolpert. I especially thank Diana Ellis, Robert Old, and Robert Spooner for commenting on the entire draft. The views expressed in this book are mine, as are any errors of fact or reasoning.

Contents

Figure Credits

While every effort has been made to trace the owners of copyright material reproduced here, the publishers would like to apologise for any omissions and would be pleased to incorporate missing acknowledgements in any future editions. In some instances we have been unable to trace the owners of copyright material and we would appreciate any information that would enable us to do so.

Frontispiece: Published by courtesy of Olympus Life Science Europa GmbH.

Fig. 1 Photo by Wolfgang Bayer Productions.

Fig. 2.3 Portrait by courtesy of the maths and computing department at St. Andrews University, UK.

Fig. 2.4 Photos by courtesy of Malin Space Science Systems Inc.

Fig. 2.11 Redrawn from Fig. 2 of G.S. Paul, *Journal of Religion and Society,* Vol. 7, 1–17, 2005. This journal is available free online.

Fig. 3.1 Wikipedia: copyright 2004 Richard Ling under the GNU Free Documentation License.

Fig. 3.2 Photos purchased from www.photolibrary.com. Fly agaric – Sven Zacek/frog – Ifa Bilderteam/ emerald tree boa – Jack Milchanowski/ orang-utan-Dale Robert Franz/three bacterial types – Dennis Kunkel/jellyfish – Mark Deeble & Victoria Stone/lesser flying squid – Marevision/ butterfly – FritzPoelking/ green-crowned brilliant humming bird – Rolf Nussbaumer/bee orchid – David Clapp,

Fig. 3.3 Copyright Don Enger (www.animalsanimals.com)

Figs. 3.5 and 4.13 Reprinted by courtesy of Mark Ridley from *Evolution*, 3rd ed., 2004, published by Blackwell.

Figs. 3.8, 3.9 and 4.16 Reprinted from *Evolutionary Analysis* by Scott Freeman and Jon C. Herron, 4e, 2007. Published by Pearson Education Inc.

Figs. 4.8 and 4.9 From *Molecular Cell Biology* by H. Lodish, A. Berk, S.L. Zipursky, P. Matsudaria D. Baltimore, & J. Darnell, Copyright 2000, 4e by W.H. Freeman and Co. Used with permission.

Fig. 4.14 Reprinted by courtesy of Nipam Patel (nipam@uclink.berkeley.edu) from *Evolution* by N.H. Barton, D.E.G. Briggs, J.A. Eisen, D.B. Goldstein & N.H. Patel, 2007, published by Cold Springer Harbor Press, W.H. Freeman.

Fig. 4.18 Reprinted by courtesy of Museum fur Naturkunde (Berlin), Raimond Spekking/Wikipedia; Copyright Pearson Education Inc.: American Museum of Natural History in New York, Matt Martyniuk (Dinoguy2)/Wikipedia.

Fig. 4.19 Reprinted courtesy of Macmillan Publishers Ltd: *Nature* April 5, 2006: Paleontology; 'A firm step from water to land' by E. Ahlberg and J.A.Clack, p. 747–749, Copyright 2006.

Fig. 4.23 Appendix – K. Kardong et al., *Vertebrates* 3e. Copyright 2002, reprinted by permission of McGraw-Hill Companies. Human embryonic tail – Wikipedia Picture of the Day, March 21st 2007. Photocredit Ed. Uthman M.D. Released into public domain. Whale – Reprinted courtesy of Macmillan Publishers Ltd, from 'Walking with Whales' by C. De Muizon, *Nature* 413, 259, 2001. Whale limb – Milwaukee Public Museum

Fig. 5.1 Eye – reprinted from *Human Biology GSCE* 3e by Morton Jenkins 1989, published by Charles Letts and Co. Retina – Reprinted from *Basic Anatomy and Physiology* 4e, 1999, published with permission by Hodder-Murray.

Fig. 5.2 Euglena – Wikipedia: copyright Shazz under the GNU Free Documentation Licence. Tripedalia – reprinted with permission from K. Nordstrom, R. Wallen, J. Seymour & D. Nilsson, *Proc, R. Soc. London B* 270, 2349–2354, 2003. Platynereis – reprinted courtesy of Macmillan Publishers Ltd: *Nature* 456, 395–399, 2008, 'Mechanism of phototaxis in marine zooplankton', by G. Jekely, J. Colombelli, H. Hausen, K. Guy, E. Stelzer, F. Nedelec, & D. Arendt. Schmidtea – reprinted from *Animal Eyes* by M.F. Land and D-E. Nilsson 2002, published by Oxford University Press.

Fig. 5.3 Reprinted with permission from AAAS from 'Casting a genetic light on the evolution of eyes' by R.D. Fernald, *Science* 313, 1914–1918, 2006.

Fig. 5.4 Wikipedia: copyright Matticus78, released under the GNU Free Documentation License.

Fig. 5.5 Normal and eyeless mutants: reprinted by permission of Walter J. Gehring, Biozentrum, Basle University. Ectopic eyes: Reprinted with permission from 'The genetic control of eye development and its implications for the evolution of the various eye-types' by Walter J. Gehring, *International Journal of Developmental Biology* 46, 65–73, 2002. Mouse embryos: Reprinted with permission of Macmillan Publishers Ltd. from *Nature* 354, 522–525, 1991, 'Mouse *small eye* results from mutations in a paired-like homeobox-containing gene' by R.E. Hill, J. Favor, B.L.M. Hogan, C.C.T. Ton, G.F. Saunders, I.M. Hanson, J. Prosser, T. Jordan, N.D. Hastie & V. van Heyningen.

Introduction: The Aims of This Book

Figure 1 summarises the two connected aims of this book. The first aim is to explain the importance of evolutionary theory in biology and the second aim is to explain how science works.

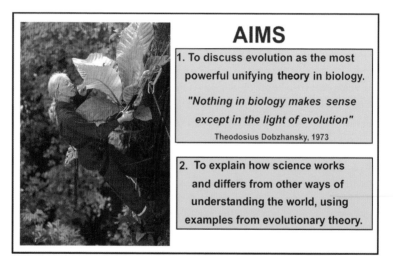

AIMS

1. To discuss evolution as the most powerful unifying **theory** in biology.

"Nothing in biology makes sense except in the light of evolution"
Theodosius Dobzhansky, 1973

2. To explain how science works and differs from other ways of understanding the world, using examples from evolutionary theory.

Fig. 1

All the branches of science today are founded on theories. Now you have to be very careful when using the word "theory" because it has two, quite different, meanings, and these are often confused with each other. In ordinary conversation, the word "theory" is used to mean a "wild or fanciful speculation, a hunch". This is what it means when, for instance, people say "One of my pet theories is that, when I am in a hurry, all the traffic lights are red".

This is not what the word **"theory"** means in science. *In science, theories are coherent conceptual frameworks that strive to unify and make sense of the maximum amount of currently available evidence in a given field.* All the modern branches of science are founded on theories. Chemistry is founded on one theory – the theory that matter is made of atoms. Geology is founded on the theory of plate tectonics

– the idea that the surface of the Earth is divided into a series of moving plates. Physics is founded on two theories – quantum mechanics that deals with the very small, and relativity that deals with the very large. Biology is, like chemistry and geology, founded on one theory, evolutionary theory, the idea that all organisms are related to one another and change over time.

It is difficult to overstate the importance of evolutionary theory for understanding the living world – evolutionary considerations permeate our understanding of biology at all levels, from ecosystems to individual organisms, from anatomy to molecules.

This is because every organism in the world is the product of an evolutionary process. This fact is summarised by a famous statement made in 1973 by a biologist with the splendid name of Theodosius Dobzhansky – "Nothing in biology makes sense except in the light of evolution". Not only is evolutionary theory fundamental to understanding the biological world, but in addition, it can plausibly be argued to be the most prominent and far-reaching theory in the whole of science. This is a strong claim, so we musk ask, why is this? It is because the idea of evolution has changed the way in which we view the place of humanity in the universe.

The idea of evolution can be traced back to the Ancient Greeks, but before it was substantiated by Charles Darwin in the 19th century, the prevailing view in the Western world was that we live in a universe specifically created for us a few thousand years ago by a benevolent god – a human-centred universe. However, discoveries of fossils of extinct organisms, the growing realisation by geologists that the Earth is much older than previously thought, and worst of all, that humanity is of very recent origin, were starting to raise doubts among some religious people even before Darwin made his contribution. An example of this doubt is provided by the poem *In Memoriam* by Tennyson, published in 1850. In this poem, Tennyson expresses his fear and dismay that the death of his friend, Arthur Hallam, at the early age of 22, was a meaningless event in a purposeless universe, and that human life has no more significance than the life of any other animal.

The extensive evidence published in Darwin's book *On the Origin of Species* in 1859 supported an alternative view to the human-centred one, in which humans are seen as one animal amongst many, created by a natural process that has no design, foresight or purpose, and that has been operating for many millions of years. In his later book entitled *The Descent of Man* (1871), Darwin further argued that humans are different from other animals only in terms of degree, not in kind. This view contrasts with the older view that human nature can be explained only on the basis of the actions and intentions of a supernatural creator. The implications of this new outlook are still stimulating vigorous debate, but among biologists, evolution is regarded as a fact as well as a theory because there are no seriously discussed alternative theories. In the same way, chemists discussing atomic theory agree that it is a fact that matter is made of atoms because there are no alternative theories.

Some of the people who kindly commented on drafts of this book expressed the fear that I am asking people to choose between belief in God and belief in evolution. This is not the case because these two subjects are not logically connected. The choice is not between God and evolution but between whether God exists or not, and whether evolution has occurred or not. Although the mainstream Christian churches opposed the idea of evolution in the past, based on a literal interpretation of the Bible, they do not do so today, and have managed to reconcile evolution with their belief in the supernatural. What I am asking people to do is to make their own assessment of the best quality, empirical evidence for each proposition, and come to their own conclusions in the light of this evidence rather than uncritically accepting dogmatic claims.

The second aim of this course is quite different – it is to explain how science works. If you listen to discussions or read articles about science in the popular media, it is obvious that there is widespread misunderstanding of how science is conducted and the type of reasoning that it uses. Interviewers, and even some scientists, talk about "proving" their ideas, but as I explain in Chapter 2, proof is not a concept applicable to science. Another common error is to use the word "theory" when what is really meant is "hypothesis"; scientists are commonly guilty of this mistake. But the most common and serious mistake is to assume that scientific conclusions are necessarily certain – this is to misunderstand the basic way that science operates. Teachers are partly to blame for all this confusion. Professors spend too much time talking about the results, the facts and theories of science, and not enough time explaining how science works as a discipline, especially how it differs from other ways of explaining the world.

In Chapter 1 of this book, I describe and compare the two principal ways in which humans try to understand the world. In Chapter 2, I discuss the type of thinking that characterises how science works, especially those features that distinguish science from religion. I describe in Chapter 3 the theory of evolution by natural selection, as formulated by Charles Darwin in the 19th century, and elaborated by geneticists, biochemists and mathematicians since his time. Chapter 4 contains some of the evidence that supports the idea that all organisms, including humans, are related to one another and change over time – the idea of evolution. As a striking example of evolution in action, Chapter 5 discusses the evolution of eyes. At the end of each chapter, you will find a Further Reading list, so that you can pursue particular topics in further depth. Some references occur in more than one Further Reading list because they discuss aspects of more than one chapter. The book ends with a list of Definitions and Suggestions for Discussion Topics.

Chapter 1
Two Ways of Explaining the World

Human beings have an innate tendency to seek explanations about the nature of the world. This tendency is most obviously seen in the curiosity of young children, who drive their parents frantic by endlessly asking questions about the world, many of which the parents cannot answer. I recall asking my parents why the sky is blue, and they were unable to tell me.

Historically, there have been two quite distinct ways of explaining the world – recall that, by "the world" in this book, I mean everything that we experience. Philosophers call these two ways, *supernaturalism* and *naturalism* (Fig. 1.1).

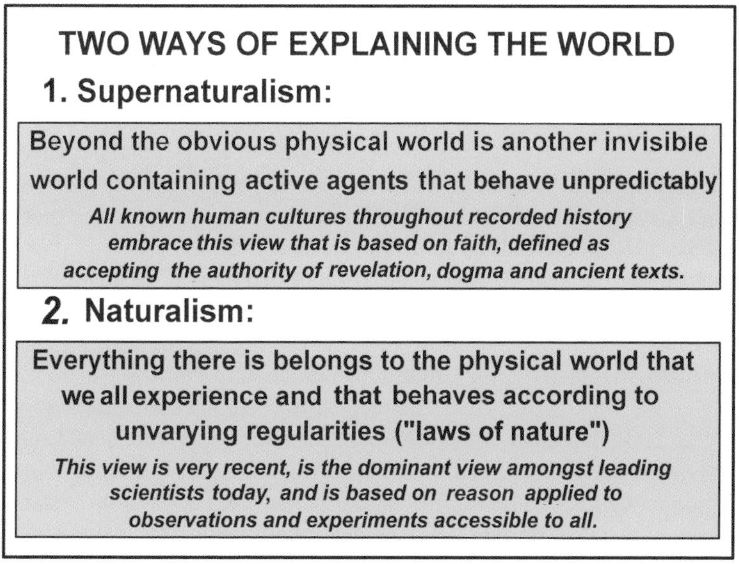

TWO WAYS OF EXPLAINING THE WORLD

1. Supernaturalism:

Beyond the obvious physical world is another invisible world containing active agents that behave unpredictably

All known human cultures throughout recorded history embrace this view that is based on faith, defined as accepting the authority of revelation, dogma and ancient texts.

2. Naturalism:

Everything there is belongs to the physical world that we all experience and that behaves according to unvarying regularities ("laws of nature")

This view is very recent, is the dominant view amongst leading scientists today, and is based on reason applied to observations and experiments accessible to all.

Fig. 1.1

J. Ellis, *How Science Works: Evolution*, DOI 10.1007/978-90-481-3183-9_1,
© Springer Science+Business Media B.V. 2010

Supernaturalism and Naturalism

Supernaturalism embraces all those ideas that suppose that, alongside the physical world that we are all aware of, there co-exists another world that is invisible, but which contains active agents, variously termed gods, deities, spirits, souls, ghosts, demons, fairies and so on. These agents are often believed to have their own agendas, their own views, preferences and purposes. They can behave unpredictably, that is, they may be capricious, and often, but not always, they are supposed to interact with the physical world.

Explaining the world in supernatural terms is extremely popular. All human cultures throughout recorded history have produced such beliefs and the vast majority of people in the world today adhere to one or other of them. The term "belief" is defined as a statement of faith that an idea is true or important, whether or not there is any testable evidence for it. The term "religion" is defined as the belief in some superhuman controlling power or powers, often requiring obedience, respect and worship. Every culture has a belief in an invisible world that contains one or more gods who are in control of powerful forces, and thus can be prayed to for advice, comfort and practical action. If we ask how this type of thinking is maintained from generation to generation, it is by accepting the authority of tradition, personal revelation and ancient texts. These sources of authority are usually regarded as sacrosanct and not open to question, but can be subject to different interpretations. Surveys show that the best predictor of peoples' religious belief is that of their parents – children tend to believe what their parents tell them because their parents are their first sources of authority.

The alternative way of explaining the world is called "naturalism" by philosophers. This view argues that there is only one world; it is the physical world we are all aware of, and it behaves according to *inbuilt, unvarying regularities* as determined by observation and experiment. These regularities are sometimes called "laws of nature" but this term is often misunderstood to imply a lawgiver, which is not the intention. Examples of such unvarying regularities are the three laws of motion proposed by Isaac Newton, and the four laws of thermodynamics. In this naturalistic scheme of things, there are no supernatural agents, so there is no possibility of miracles nor is there any point in praying for divine intervention, other than to make people feel better.

The naturalistic view is very recent in human history. It began in a serious way only at the time of the Enlightenment in the eighteenth century in Western Europe. The Enlightenment is the term used to describe an intellectual movement whose members believed that reason could be used to combat both superstition and tyranny and to build a better world. The principal targets of the founders of the Enlightenment were organised religion and the domination of society by an hereditary aristocracy. The founders of the Enlightenment were motivated by the desire to be free to pursue the truth as they saw fit, without the threat of sanction for challenging established ideas.

Naturalism argues that reason should be the prime means of understanding the world, but reason based on observation and experiments and not on an acceptance of ancient authority. Reason is defined as the intellectual faculty by which conclusions are drawn from premises. Now you might say that I am an ancient authority – I am certainly ancient! – and that you are accepting what I say because I am an authority, but the difference is that any scientific claim I make can be checked by anybody prepared to take the time and trouble, whereas it is not possible to check claims derived from revelation or ancient texts – you either accept such claims or you don't.

The Incompatibility of Religion and Science

Now it is important to understand that *both naturalism and supernaturalism are assumptions, and both are logically possible, but both cannot be correct* – by definition, one excludes the other. This exclusion arises because the word "supernatural" describes by definition a hypothetical realm which cannot be observed or recorded by the procedures of science. Supernatural agents by definition posssess properties above and beyond the natural world and its properties – that is why we use the word "supernatural". So supernatural agents are not constrained by the unvarying regularities implicit in the naturalistic assumption. If we could apply natural knowledge to understand supernatural agents, then by definition they would not be supernatural.

It follows that science is incompatible with religion. Why is this? It is because once you attribute any particular event to a supernatural agent, a proposition that cannot be disproven by observation or experiment, then science becomes both irrelevant and impossible. This is because science works on *the assumption that natural events have natural causes.* For example, if a scientist carries out an experiment and finds that he or she cannot initially understand the results of that experiment, the scientist does not say that is because of the actions of supernatural agents – if a scientist did say that, science would stop, because the actions of supernatural agents by definition are not subject to unvarying regularities. What scientists do instead, is to think more imaginatively about the problem, until they come up with another testable hypothesis involving natural causes.

I will address in Chapter 2 the reasons why some scientists nevertheless hold religious beliefs, but it is important to note here that even religious scientists do not introduce supernatural explanations into their science. If they did, then anything that is logically possible might become actual, despite the unvarying regularities that characterise the natural world. It follows that introducing religious explanations into science would destroy the practice of science. So supernaturalism is not included within science because, by its very nature, it is not testable. Supernaturalism *lacks a methodology* by which its claims can be tested, whereas science does have such a methodology. How this methodology operates I discuss in Chapter 2.

The naturalistic viewpoint that defines science is more accurately termed "*methodological naturalism*" because of its emphasis on methodically testing ideas. This term is used to distinguish this type of naturalism from a separate type, called by various names – *ontological* naturalism, or *philosophical* naturalism, or *metaphysical* naturalism. These three names are equivalent, and I shall henceforth use the term metaphysical naturalism. The word "metaphysical" means literally "beyond physics", and refers to those ideas that suppose the existence of a world above and beyond the physical world that we are all aware of. Metaphysical naturalism states that the supernatural does not exist. Thus, unlike methodological naturalism, which is an *assumption*, metaphysical naturalism is an *assertion*. It is an assertion that is incapable of being tested, because it states a negative proposition, and it is not logically possible to disprove a negative proposition. So metaphysical naturalism is not part of science, but of course it is a view that can be held by scientists, as well as by anyone else. In the Further Reading at the end of Chapter 2, I have listed articles by two philosophers, Barbara Forrest and Robert Pennock, who addressed these issues at the Dover court trial in Pennsylvania referred to in the Preface.

I emphasise this difference between supernaturalism and methodological naturalism because some people, including some scientists, suggest that there is no conflict between the two – that science and religion are compatible because they have different aims and deal with different areas of human experience. Such people argue that science tries to discover what things are and how they work, while religion is trying to discover whether the Universe, and human life in particular, has any overall meaning. However, this view that science and religion concern different areas of enquiry is both logically and historically incorrect.

It is logically incorrect because both science and religion have exactly the same aim – to understand the world and our place in it. It is historically incorrect because throughout recorded history, religions have tried to answer questions about what things are made of and how they work as an essential part of their mission, just as science does. All religious beliefs contain creation narratives about how the world originated in the physical sense. What has changed over the time since science started to develop in the seventeenth century, is that many religions have progressively abandoned trying to explain how the world works, as the creation myths were progressively shown to be unsupported by the physical evidence. So some mainstream religions today concentrate instead on whether the world has any purpose or moral dimension, and no longer claim to study how the world works.

The late evolutionary biologist, Steven Jay Gould, proposed in 1997 the idea that science and religion are concerned with different domains of understanding that he termed "magisteria". "Magisterium" is the Latin word for "teacher". This view is summarised by the acronym NOMA, for "non-overlapping magisteria". According to the NOMA proposal, science is concerned with what the Universe is made of and how it works, while religion is concerned with questions of ultimate meaning and moral value. Because of this difference, Gould argued, religion and science cannot be combined and are not in conflict. They deal with different areas of human experience, so it follows that science and religion cannot comment on each other's concerns.

The problem with this NOMA idea is that there is no empirical evidence as yet that the Universe has any overall meaning or moral dimension, and Gould does not attempt to offer any. The main-stream religions derive their sense of meaning and moral purpose from supernatural sources, which leaves them with no room to manoeuvre to accommodate scientific discoveries about the nature of human behaviour. NOMA also implies that religious people should not try to reinterpret their beliefs in the light of scientific discoveries, which is unfair to them. It is very clear that religion cannot, and does not, refrain from making claims that have observable consequences in the physical world, such as the occurrence of miracles and the answering of prayers, so it is untrue that religion and science do not overlap. The fact that many sick people visit Lourdes every year in the hope of miraculous cures is convincing evidence of that untruth. Nevertheless, some scientists accept the NOMA principle. When doing their science, they accept the naturalistic assumption that the supernatural does not exist – if they did not, they could not practise science, as I explain above. When practising their religion, they abandon the naturalistic assumption. In this way, they enjoy having their cake and eating it. Robert Old has suggested that this practice by religious scientists be termed *occasional theism.*

Historically, methodological naturalism has been found to be the best way we have so far discovered to make any progress in understanding the world – it is naturalism, and not supernaturalism, that has made the modern developed world. In this world we find ourselves in the paradoxical situation that our advanced lifestyles have been created by the application of scientific discoveries, but despite this, the vast majority of people enjoying those lifestyles still interpret the world in a supernatural fashion.

There are several possible reasons for this. One is that the evolutionary process that determined how the human mind works predisposes us to interpret the world in a supernatural fashion; I discuss this possibility later in this book. Another reason is that most people, including some scientists, have not been educated to understand the basic principles by which science operates, so in Chapter 2, I describe these principles.

Further Reading

1. Wikipedia articles provide useful, but non-peer reviewed, introductory Discussions of: *Naturalism (Philosophy)/Scientific Method/History of Science/ Origin of Religion/Evolutionary Psychology of Religion.*
2. *Methodological Naturalism and Philosophical Naturalism: Clarifying the Connection*: Barbara Forrest. http://www.infidels.org/library/modern/barbara_forrest/naturalism.html. This article discusses the philosophical basis of naturalism.
3. *Expert Report at the Kitzmiller Dover Trial:* Robert T. Pennock. http://www.msu.edu/~ pennock5/ Scroll down to 'expert report'. A philosopher of science explains how science works and differs from religion, and why intelligent design is not a scientific concept.

Chapter 2
How Science Works

There are three features of science that distinguish it from supernaturalism. I call these features: *nullius in verba*, Occam's razor and uncertainty

Nullius in verba

The first distinctive feature is shown in Fig. 2.1.

THE DISTINCTIVE ASPECTS OF SCIENCE

1.*"Nullius in verba"* - the motto of the Royal Society

Translation - "not in words"

Or: Do not take anyone`s word for it

THE ROYAL SOCIETY

NATIONAL ACADEMIES - national clubs of leading scientists

UK - The Royal Society: 1400 Fellows (44 elected each year)

USA - The National Academy of Sciences: 2000 Members,

(72 elected each year)

Fellows & Members are elected on the basis of the quality of their *peer-reviewed* papers in the primary scientific literature

Fig. 2.1

J. Ellis, *How Science Works: Evolution*, DOI 10.1007/978-90-481-3183-9_2,
© Springer Science+Business Media B.V. 2010

"*Nullius in verba*" is the motto of the Royal Society, the premier body of scientists in Britain, founded in 1660. Other countries have similar bodies of leading scientists, called National Academies. The literal translation of this motto is "not in words". This motto encapsulates the view that you should base your beliefs on your own assessment of the best available evidence and not take anyone's word for it. So there is no room for dogma or for an appeal to authority, tradition or ancient texts – this is a major difference from how supernaturalism operates. This difference between supernaturalism and naturalism can be summarised by saying that, while supernaturalism has authorities who make assertions, naturalism has experts who are familiar with the best available evidence.

But now we have to ask – where is the best available evidence to be found? The answer is that the best available evidence is available in the peer-reviewed literature. Figure 2.2 explains what this means. When a scientist, or more likely these days, a group of scientists, feel that they have made some new observations or arrived at

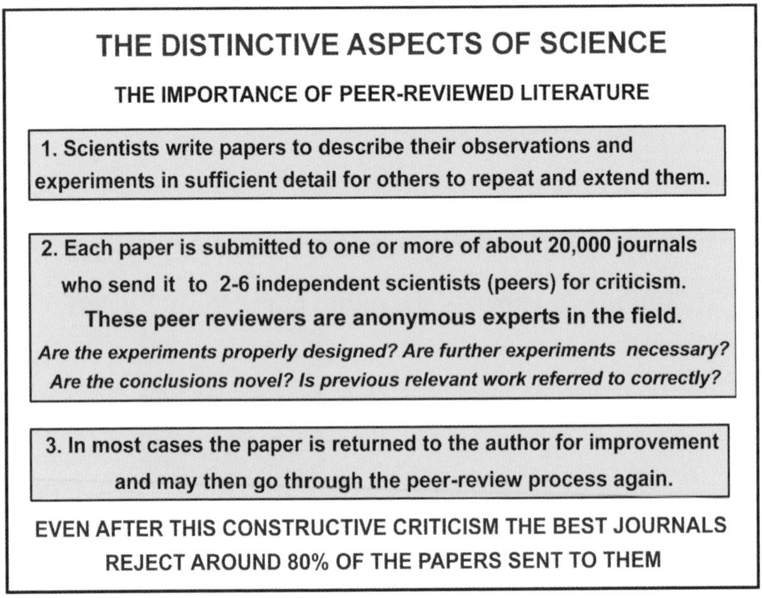

THE DISTINCTIVE ASPECTS OF SCIENCE

THE IMPORTANCE OF PEER-REVIEWED LITERATURE

1. Scientists write papers to describe their observations and experiments in sufficient detail for others to repeat and extend them.

2. Each paper is submitted to one or more of about 20,000 journals who send it to 2-6 independent scientists (peers) for criticism. These peer reviewers are anonymous experts in the field.
Are the experiments properly designed? Are further experiments necessary? Are the conclusions novel? Is previous relevant work referred to correctly?

3. In most cases the paper is returned to the author for improvement and may then go through the peer-review process again.

EVEN AFTER THIS CONSTRUCTIVE CRITICISM THE BEST JOURNALS REJECT AROUND 80% OF THE PAPERS SENT TO THEM

Fig. 2.2

some novel insights about some aspect of the natural world from their experiments, they write down what they have done in enough detail for other scientists familiar with the field to be able to repeat the observations and experiments.

The main reason for doing this is to ensure that the observations are reliable, in the sense that they can be repeated by other, independent scientists. The completed writing is referred to as a manuscript or paper, and this is submitted to a learned journal that specialises in the appropriate branch of science. The journal editor sends the paper out for review by several anonymous experts in the field – the peers. The term "peer" means "equal", but is often misunderstood to mean "superior"

These peers are asked to read the paper in detail and to assess the validity of the observations, experiments and conclusions. They may suggest that some of the conclusions are not justified until further observations or experiments have been performed or point out flaws in the reasoning used by the authors. They may think that, although the conclusions are valid, they are not sufficiently important or novel to justify publication in the journal to which they have been submitted. The editor passes on these criticisms to the authors and asks them to revise the paper in light of the peers' comments. Such is the pressure on space in the best journals that only those papers that contain the most innovative observations can be accepted for publication – the others are rejected and the authors then may send them to other journals, where the entire process is repeated.

Once a paper has been accepted into the peer-reviewed literature, its assessment by other scientists does not stop but continues. Its conclusions are critically discussed at science conferences and in laboratories around the world. Eventually a general consensus on a given topic emerges. An example would be the general consensus that global warming has a human-made component. This does not mean that every climate scientist agrees with this conclusion, but it does mean that the weight of the evidence available today points in this direction. Future discoveries may of course modify this conclusion. *Scientific ideas are always open to challenge and change in the light of new evidence.*

Contrast this elaborate assessment procedure with the lack of such procedures by which religious claims can be assessed. How can you assess claims made in documents written hundreds or thousands of years ago by people whose knowledge of the world was inferior to ours? You either accept such claims based on the authority of the person making it, or you do not. What you cannot do is to assess such claims in the same way that you can assess scientific claims by reference to the peer-reviewed literature. This does not mean that all the claims made in the peer-reviewed literature are correct – scientists are human, they make mistakes, they are prone to dogma, and are influenced by things such as politics, seniority, charisma and one-upmanship – but the peer-reviewed literature is still the best source of information about the world that we have. The same is true for the arts and the humanities – the peer-reviewed literature in these fields of study is the best source of information about the state of understanding in these fields, which include history, philosophy and theology. You should not accept at face value any reports in the media about scientific matters unless and until they have appeared in the peer-reviewed literature.

Occam's Razor

The second distinctive aspect of science, and in my view, the most important, is illustrated in Fig. 2.3.

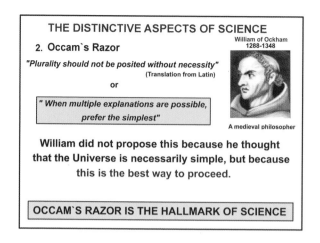

Fig. 2.3

William of Occam was a Franciscan philosopher who came from the village of Ockham (or Occam) in the county of Surrey in the United Kingdom. He tackled the problem that for any given body of evidence you can almost always postulate several, quite different explanations. William argued that in this situation the best way to proceed is to prefer the simplest explanation that is consistent with all the available evidence – the explanation that makes the least number of assumptions. The word "razor" is used to mean that unnecessary assumptions are shaved away. Occam's razor is sometimes referred to as the "principle of parsimony"; the word "parsimony" means "economy". Let me give you an example that actually happened recently.

In 1976 the Viking spacecraft took the photograph of a rock formation on the Martian surface, shown in Fig. 2.4a. This photograph caused a lot of excitement because many people, including some scientists, interpreted it to mean that Martians had carved the image of a human face on the rock. Now this interpretation clearly makes a lot of assumptions – that intelligent creatures exist or have existed on the planet Mars, that they know what humans look like, or even look like humans

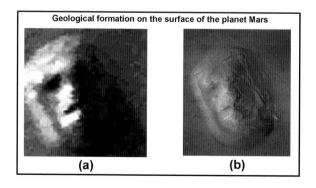

Fig. 2.4

themselves and that they want to signal to us. A much simpler interpretation, that is one with fewer assumptions, is that this is just an accidental effect of the angle of sunlight that happens to remind us of a face. We humans are programmed to recognise faces. Some people see faces in clouds, fires, teacups and items of pastry!

The picture shown in Fig. 2.4b was taken in 1998 by the Global Surveyor spacecraft when the angle of illumination of the same region was different, and it clearly supports the simpler interpretation. But let's be honest – the simpler interpretation is also more boring! The reason that conspiracy theories are so popular is that they are often more interesting than real life.

Now it is important to grasp that Occam's razor does not say that you should prefer the simplest hypothesis because it is more likely to be correct – there is no *a priori* reason why Nature should be simple. What Occam's razor says is that *you should prefer the simplest hypothesis because it is the best way to proceed.* So William is defining a method – an essential part of the scientific method. It is one of the ironies of history that William did not apply his razor to his religious beliefs.

I cannot overemphasise the importance of Occam's razor to the practice of science. If you abandon this principle, you might as well believe any interpretation of the world that you find comforting and appealing – and many people do. It is probably because of Occam's razor that the majority of leading scientists today are not supernaturalists, because postulating invisible active agents clearly requires more assumptions than does the naturalistic view that such agents do not exist. Let us now look at some of the evidence about religious belief amongst scientists.

Religious Belief amongst Leading Scientists

A survey of the incidence of religious belief among leading American scientists was published in the journal *Nature* in 1998, and Fig. 2.5 compares it with similar surveys in 1914 and 1933. The response rate in the 1998 survey of members of the US National Academy of Sciences was about 50%.

SCIENTISTS AND SUPERNATURALISM

THE INTERNATIONAL JOURNAL OF SCIENCE

nature

Vol. 394, JULY 23, p.313, 1998

correspondence

" Leading scientists still reject God "

Survey of leading American scientists

	1914	1933	1998
Belief in a personal God	27.7 %	15%	7.0%
Disbelief in a personal God	52.7%	68%	72.2%
Doubt or agnosticism	20.9%	17%	20.8%

Fig. 2.5

These surveys clearly show a decline of declared belief in a personal God during that period. A similar unpublished survey was carried out among Fellows of the Royal Society in Britain in 2006 and obtained a similar result – the incidence of supernaturalists is less than 10% among those Fellows who responded to the survey (about 25% of all the Fellows contacted). But interestingly, this is a recent trend, and if we look further back in the history of science, we see a different pattern.

We need first to ask the question – who was the first scientist in the modern sense?

What I mean by this is – who is the first person we know about who used the basic experimental methodology that modern scientists use? Figure 2.6 provides the answer and lists the basic steps in the experimental method.

SCIENTISTS AND SUPERNATURALISM
QUESTION: Who was the first scientist?

THE METHOD OF EXPERIMENTAL SCIENCE

1. Ask questions about some aspect of the Universe.

2. Imagine a possible explanation i.e. formulate a hypothesis.

3. Test the hypothesis with a suitable experiment.

4. Modify the hypothesis to accommodate the results

5. Perform more experiments to test the modified hypothesis.

ANSWER: Ibn al-Haytham (Alhazen)

An Arab physicist of the 10th century (965-1040), he was part of the Arabic Golden Age that lasted from the 8th to the 13th century

LIKE ALHAZEN, MOST LEADING SCIENTISTS FROM GALILEO, NEWTON & FARADAY TO BOHR, HEISENBERG & PLANCK, WERE
SUPERNATURALISTS

Fig. 2.6

To be a scientist you have to be curious about the Universe and ask questions about aspects that you do not understand. You have to concentrate on a particular problem that looks to be of a manageable size, and imagine possible explanations. You then devise experiments to test these explanations, and ask whether the natural world behaves in accordance with one or other of your ideas. You usually have to modify your ideas to accommodate the results of the experiments, using Occam's razor as a guide.

This procedure is not foolproof nor is it guaranteed to give you the correct answers. Occam's razor can mislead you if, for example, the correct explanation is not among the possible explanations you have imagined. Thus the methodology of science is not a surefire recipe; its successful use requires not only intelligence, but also creativity and imagination, and the honesty to admit mistakes when you make them.

To return to the question as to who was the first experimental scientist that we know about, the answer is surprising. Most people think it was an Ancient Greek philosopher, such as Aristotle, Plato or Archimedes. Ancient Greek philosophers based their views in many cases on empirical observations of the world rather than by appeals to authority, and thereby made many advances in understanding, but they were not renowned for formulating hypotheses and then testing them by experiment in the routine way that modern scientists do.

The first recorded scientist, defined as I have described, was an Arab called Ibn al-Haytham, who carried out experiments on many aspects of vision and optics in the tenth century (Fig. 2.6). This was the middle of a period when in the Arab world there was a tremendous flowering in the arts, literature and sciences, called the Arabic Golden Age. Some people call it the Islamic Golden Age but many people of different faiths contributed to it, and they all wrote in Arabic. It is also surprising to learn that many of the discoveries that we in the West associate with people like Copernicus, Galileo and Newton in the 16th and 17th centuries were initiated in the Arab world, at a time when free enquiry was not encouraged by the religious authorities in Europe. Ibn al-Haytham was the first person to demonstrate that light travels from objects into eyes and not from eyes onto objects, as had been suggested by earlier Greek philosophers. He also showed by experiment that light travels in straight lines.

Now Ibn al-Haytham was a devout Muslim – that is, he was a supernaturalist. He studied science because he considered that by doing this he could better understand the nature of the god that he believed in – he thought that a supernatural agent had created the laws of nature. The same is true of virtually all the leading scientists in the Western world, such as Galileo and Newton, who lived after al-Haytham, until about the middle of the twentieth century. There were a few exceptions – Pierre Laplace, Simeon Poisson, Albert Einstein, Paul Dirac and Marie Curie were naturalists for example. Charles Darwin experienced a decline in his Anglican belief in a benevolent god as he grew older, and in a private letter written two years before his death in 1882 said "I have never been an atheist in the sense of denying the existence of God... I think that generally (and more and more so as I grow older), but not always... that an agnostic would be a more correct description of my state of mind". The term "agnostic" was coined in 1869 by Darwin's friend and supporter, Thomas Henry Huxley, who defined it as the position that "it is wrong for a man to say that he is certain of the objective truth of a proposition unless he can provide evidence which logically justifies that certainty". In other words, Huxley was disputing the statement by religious people that their knowledge of the supernatural is certain.

Both agnosticism and atheism come in strong and weak versions. The strong agnostic thinks that the supernatural is unknowable even in principle, while the weak agnostic thinks that there is no empirical evidence to support the existence of the supernatural, and that therefore one should not believe in the supernatural unless, and until, such evidence is found. The strong form of atheism is defined as the belief that the supernatural does not exist, while the weak form of atheism is indistinguishable from the weak form of agnosticism. One way to remember the difference between weak agnosticism and strong atheism is that weak agnosticism *assumes* that the supernatural design does not exist because there is no evidence for it, but does not rule it out as a logical possibility, while strong atheism *asserts* that the supernatural does not exist and does rule it out as a logical possibility. This is a debate where definitions are important. I shall emphasise the necessity of defining terms exactly in a later section.

Other correspondence suggests that Darwin was sometimes inclined to the deist view that an intelligent agent had created the Universe and the laws by which it operates, but thereafter had no interaction with it. This deist view is equivalent in practical terms to being a naturalist because it allows science to function in a naturalistic framework, as well as ruling out miracles and praying for divine intervention. For this reason, deism is the only type of religious belief that is not in conflict with the naturalist approach upon which science depends. Most of the other leading scientists up to about the middle of the twentieth century were supernaturalists however.

What can we deduce from these historical facts? Firstly, it is obvious that believing in the supernatural does not prevent you becoming a leading scientist. We can also deduce that such people separate the way they think about science from the way they think about religion. When doing science they use Occam's razor, but when doing religion they abandon Occam's razor, because postulating invisible agents clearly requires more assumptions than not postulating them. These assumptions include the origins, properties, and interests of such agents. These assumptions vary greatly among the different religions – in the monotheistic religions God is good, omniscient and omnipotent (hence the capital G), but this is not the case for some of the gods in many polytheistic religions. So the worst that supernaturalist scientists can be accused of is inconsistency.

What about the evidence that, according to polls and surveys, most leading scientists today have no religious beliefs? One can only speculate about the reasons for this. My suggestion is that it is because, at the end of the day, religious explanations are not really explanations – they may be emotionally appealing but they are intellectually unsatisfying, because they posit even greater mysteries that the ones you are trying to explain. As the philosopher Anthony Grayling so eloquently puts it "To answer the question of how the universe came into existence by saying "God created it" is not in fact to answer the question, but to explain one mystery by appealing to an even greater mystery – exactly like saying that the universe rests on the back of a turtle, and then ignoring the question of what the turtle rests on".

One interesting problem that some naturalist scientists and philosophers of today are tackling is how to explain the universal persistence of supernatural beliefs – why do they occur, what accounts for their particular features, what purposes do they serve? In recent years, anthropologists studying the huge variety of supernatural beliefs found around the world, and psychologists seeking evolutionary explanations for religious beliefs, have proposed a number of hypotheses (Fig. 2.7).

There are two general types of explanation offered, but they are not mutually exclusive – elements of both may be correct. The first type assumes that supernatural beliefs have direct survival value for humans and thus are adaptive features.

The obvious, common sense purposes listed in Fig. 2.7a are certainly found in the world's major religions and help people cope with suffering, especially illness. Before the real causes of human disease started to be identified some 150 years ago, people often interpreted illness as punishment for flouting the will of a deity, so they would plead with their deity for relief. Humans are also probably the only species that are aware they are going to die and they often fear what may lie beyond. Most people grieve when they lose their loved ones and like to believe that their personalities survive in some sense after death. However, anthropologists such as Pascal Boyer point out that there are many thousands of minor supernatural beliefs where these obvious, common sense explanations do not always apply, so they suppose there must be some deeper underlying reasons for supernatural beliefs, related to how the cognitive systems in the human brain interpret the world.

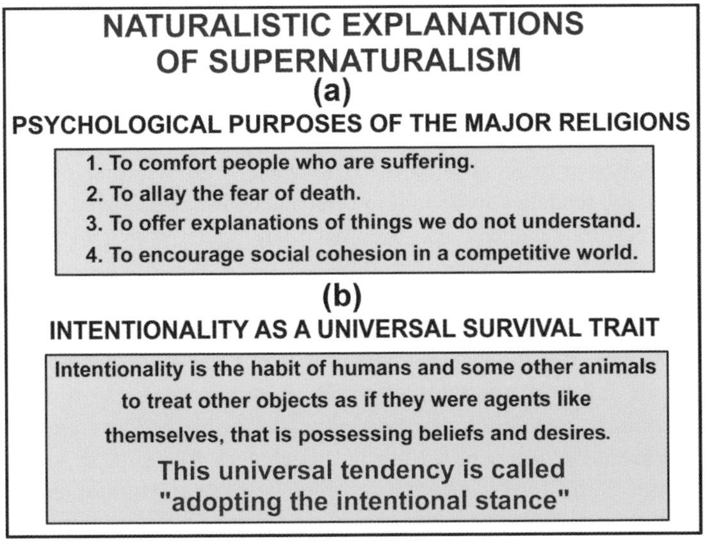

NATURALISTIC EXPLANATIONS OF SUPERNATURALISM
(a)
PSYCHOLOGICAL PURPOSES OF THE MAJOR RELIGIONS

1. To comfort people who are suffering.
2. To allay the fear of death.
3. To offer explanations of things we do not understand.
4. To encourage social cohesion in a competitive world.

(b)
INTENTIONALITY AS A UNIVERSAL SURVIVAL TRAIT

Intentionality is the habit of humans and some other animals to treat other objects as if they were agents like themselves, that is possessing beliefs and desires.

This universal tendency is called "adopting the intentional stance"

Fig. 2.7

One plausible suggestion is what is called intentionality, sometimes called the "theory of mind" (Fig. 2.7b).

Intentionality is the ability of humans and some other animals to treat other objects and animals as agents like themselves, that is, agents with minds that have desires, beliefs and intentions. Each time you have a conversation, you adopt an intentional stance towards the other person – you make assumptions about their desires, beliefs and intentions because you believe the other person is an active agent like yourself with a mind like yours. The other person is making similar assumptions about you. The term "adopting the intentional stance" was suggested by the philosopher Daniel Dennett.

Adopting the intentional stance is clearly an important survival tactic for animals, especially for social animals like ourselves, whose success improves if we cop-operate with other humans. So the suggestion is that our brains are so hard-wired to produce this type of thinking that we tend to extend it to other objects and events that affect us. For example, when a rock falls and injures us, many people tend to assume that this means that there is an active agent making the rock do this – they believe that the rock moves because of some intentionality. Such people commonly think that the agent is invisible because they cannot see an agent. In primitive societies today, it is believed that many objects in the environment – trees, rocks, rivers, mountains and so on – are inhabited by invisible spirits that can be influenced by ritual practices. How many among the most rational of us shout at our PCs when they do not do what we want? It is easy to slip momentarily into responding as though they were active agents. We all sympathize with Basil Fawlty losing his temper with his car when it refuses to start in the BBC TV comedy series *Fawlty Towers*. This adoption of the intentional stance is also a common experience among survivors of life-threatening accidents – they attribute meaning to their survival in terms of actions by a supernatural agent.

Evidence that this tendency to interpret the world in a supernatural fashion is partly genetically determined comes from the Minnesota twin studies, in which the religiosity of identical twins raised apart in different environments was compared with that of fraternal twins raised apart. The results were interpreted to mean that about 50% of the tendency to be religious is genetically determined. Further studies showed that this tendency also becomes more apparent as children approach adulthood.

Given the universality of interpreting the world in an intentional fashion, it is perhaps not surprising that some scientists have also fallen into this trap. For example, the first version of the Gaia hyothesis of James Lovelock defined Gaia as "a complex entity involving the Earth's biosphere, atmosphere, oceans and soil: the totality constituting a feedback or cybernetic system which seeks an optimal physical and chemical environment for life in this planet". The simpler view, held by most scientists, is that the biosphere is a comprehensible mixture of air, water, soil and organisms, whose behaviour is explicable in terms of different steady states produced by negative feedback effects. There is no sense in which such a system can be said to seek anything, and Lovelock has since stated that this aspect of his original proposal was meant only in a metaphorical sense.

Similarly, some physicists suggest from the fact that the Universe has properties that allow human life to appear, that this means that its properties were designed with this intention in mind – the so-called strong anthropic principle. It is hard to imagine a finer example of unconscious arrogance, as well as ignorance of the mechanisms of evolution, than to assert that humans are the purpose of the Universe, but this example does illustrate the tendency we all have to adopt the intentional stance.

On this intentional hypothesis, the widespread tendency to explain the world in supernatural terms is not itself an adaptive feature of evolution, but a byproduct of parts of the mind that evolved because they aid survival in other ways. It is essential to realise that a common feature of evolution is that properties which evolved because they help survival in a particular environment may have important, but quite unrelated, consequences, later on. For example, the evolution of feathers in some dinosaurs was probably connected with the development of warm-bloodedness, which requires insulation to combat the loss of heat, but had the unrelated, but important, consequence of permitting the subsequent development of flight.

Thus the universality of the intentional stance does not mean that supernatural beliefs are necessarily correct, only that they may originate naturally in the cognitive systems that all humans use to interpret the world. This hypothesis amounts to saying that, while supernaturalism tells us something about what goes on inside the human head, there is currently no convincing evidence that it also tells us what goes on outside the human head. So we have to be alert to avoid being victims of our biology, especially today when we find ourselves living with a stone-age mentality in a space-age world that is changing much more rapidly than our brains are evolving.

Uncertainty

The third distinctive feature of science is illustrated in Fig. 2.8. *The provisional nature of scientific knowledge is the aspect most commonly misunderstood by non-scientists.* Often when science is mentioned in the media, the impression is given that scientific knowledge is absolute and certain. This is not the case.

THE DISTINCTIVE ASPECTS OF SCIENCE
3. Uncertainty

Contrary to popular opinion, all scientific understanding is PROVISIONAL

SCIENCE IS A SET OF IDEAS ABOUT
HOW THE UNIVERSE WORKS
These ideas are based on the best observational and experimental data available at the time but are always open to change to accommodate new data. Thus science has an inbuilt self-correcting mechanism that accounts for its unmatched success at improving the human condition.
BUT
BECAUSE FUTURE EVIDENCE CANNOT BE PREDICTED, WE CAN NEVER BE CERTAIN THAT NEW DATA WILL NOT CHANGE EXISTING IDEAS. THUS ALL SCIENCE IS THEORY AND SCIENTIFIC THEORIES CAN NEVER BE PROVED.

Fig. 2.8

If we ask what science is, it is not a set of data, it is not a set of techniques, it is a set of ideas (Fig. 2.8). These ideas are based on reason applied to data and techniques, but data and techniques do not on their own constitute science – it is the ideas that constitute science. These ideas are based on the best evidence available at the time, but they are not sacrosanct – they are always open to change to accommodate new data and new ideas. It is this openness to change that explains why science is so successful at understanding the world.

Let me be very clear about this – *science is the most successful human endeavour in history.* Despite all the problems in the world, it is the case that never before in human history have so many people been so well fed, and have had so many opportunities to lead long, healthy and interesting lives. These advances stem from the application of scientific ideas and discoveries. *Science works*, so it may seem surprising that scientific knowledge is not certain in an absolute sense. Why is this?

It is because you cannot predict the future. You can never be certain that even long-held and very successful scientific ideas will not change as a result of future discoveries. Even if you had a theory that was completely correct, you could never be sure that this was the case, because you could never be certain that future discoveries would not falsify it. Let me give you an example:-

In 1687 Isaac Newton published his master work that marks the beginning of science in the Western world – the *Principia Mathematica*. This was the first book to propose general natural laws in a quantitative fashion. Newton's laws of motion and his equations that describe gravity are incredibly successful and precise – precise enough to be used to send astronauts to the Moon and to land spacecraft on the planet Mars. Nevertheless, the concepts on which Newton based his laws of motion were shown to be incorrect just over 200 years later by Albert Einstein in 1905. Newton's concepts with respect to motion are wrong because he supposed that time and space are absolute and independent – this agrees with our common-sense perceptions of time and space. Einstein had the genius to realise that because experiments show that the velocity of light is constant, irrespective of the velocity of the light source, this view must be wrong – time and space are relative to one another, not independent. Experiments done since Einstein's time show that the faster a clock moves, the slower it ticks and that the faster an object moves, the heavier it becomes. In other words, our common-sense perceptions of time and space are wrong. Newton's laws of motion work well enough in practice because relativistic effects become significant only at velocities much faster than the ones we normally have to deal with. These effects impact on human affairs only in the design of particle accelerators that would not work unless the relativistic effects were taken into account in their design.

So here we have an example of a very successful scientific theory that was accepted for over 200 years, but whose basic concepts, on which the Industrial Revolution was partly founded, are now known to be incorrect. *Science is a uniquely successful human activity precisely because it employs this inbuilt, self-correcting mechanism.* So certainty is an illusion. Scientific knowledge is not a fixed destination but a moving target. This is why media commentators discussing science who use the word "proof" demonstrate that they do not understand how science works. In science, proof is not an option. Disproof on the other hand *is* an option-if we discovered human fossils in rocks older than the rocks containing dinosaurs, our current ideas about the evolution of mammals would be instantly disproved. Einstein famously said that no amount of experiments could prove him correct, but a single experiment could prove him wrong. So if you crave certainty, you will not find it in science. This is why media interviewers who ask scientists "Are you certain that. . .?" are demonstrating their ignorance of how science works.

Another problem with some media discussions of scientific issues is the common failure to define terms accurately, so I will now discuss this problem.

The Importance of Defining Terms

The English philosopher Cyril Joad (1891–1953) is remembered today largely for being a public intellectual who boasted of his habit of cheating the rail companies by travelling without a ticket, a practice that led eventually to a court appearance and fine. This is a pity, because he used to be famed for responding to any question with the catchphrase "It all depends on what you mean by. . .". He did this so often that he was ridiculed for this response, but of course he was right. All our ideas about the world are expressed in words. So it is not a semantic quibble to insist on agreed definitions of terms that are used to discuss these ideas. This importance of defining terms applies to all forms of human discourse, not just to science. Many terms in common use have several different meanings, terms such as "organic", "spiritual", "intelligence" and "love", and it is not always clear what meaning is intended, so that discussions about these subjects easily become confused.

I should now like to talk about some of the key terms used in science. When I was at university I was taught the importance of defining terms in order to have sensible conversations. So if I ask a student to define the term "molar" say, or "eukaryote", or "evolution", I am hoping that they can rattle off a precise definition in one sentence for each term. I used to keep a notebook in which I wrote definitions of scientific terms and memorised them. I did not always understand these definitions when I first wrote them down, but I did find that, as my knowledge developed, being familiar with these definitions did help me understand what they mean.

Facts, Theories and Hypotheses

Figure 2.9 lists definitions of three terms used in science – facts, hypotheses and theories. These terms are often misunderstood by nonscientists. Even scientists tend to say "theory" when they really mean "hypothesis".

I talked about theories in the Introduction of this book. Let me remind you that theories are coherent conceptual frameworks that unify and make sense of all the available evidence in a given field. Good theories are quantitative and lead to experiments that uncover previously unknown phenomena. All the branches of modern science are founded on theories – in the end, science *is* theory. However Francis Crick, the codiscoverer of the structure of DNA, cautioned that one should beware of theories that explain all the relevant facts, because some of these facts will be wrong – scientists make mistakes!

What about "facts"?

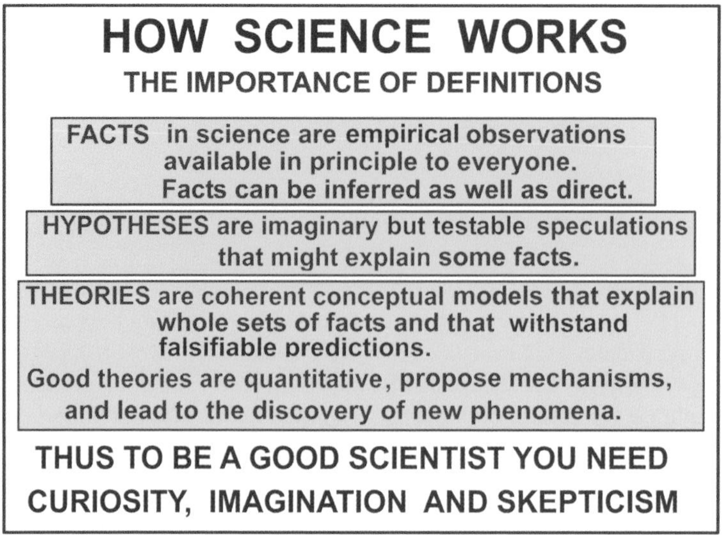

HOW SCIENCE WORKS
THE IMPORTANCE OF DEFINITIONS

FACTS in science are empirical observations available in principle to everyone.
Facts can be inferred as well as direct.

HYPOTHESES are imaginary but testable speculations that might explain some facts.

THEORIES are coherent conceptual models that explain whole sets of facts and that withstand falsifiable predictions.
Good theories are quantitative, propose mechanisms, and lead to the discovery of new phenomena.

THUS TO BE A GOOD SCIENTIST YOU NEED CURIOSITY, IMAGINATION AND SKEPTICISM

Fig. 2.9

To be considered a fact, observations must be empirical, repeatable, and shareable by everyone. The word "empirical" means derived from observation and experiment, not from what someone tells you. So if a person tells you that they had a dream in which God told them to do something, this is not a fact in scientific terms – because this observation is not repeatable by others or shareable directly with them, nor can it be disproved. This does not mean it might not be true, but it does mean that this sort of claim cannot be used in science.

Now it is important to realise that facts do not need to be based on direct observations. Many of them are of course, but facts can be also inferred from indirect observations. For example, no one has seen an electron directly, but the existence of electrons is inferred from so many indirect observations that their existence is regarded as a fact. We shall see later that some aspects of evolution cannot be observed directly because they occur far too slowly to be seen in the human lifetime, but they are inferred from so many indirect observations that they are regarded as facts.

Hypotheses are acts of creative imagination – speculations that might explain some facts. Darwin's idea of evolution by natural selection was initially proposed as a hypothesis, but the amount of empirical evidence in support of it is so large it is now regarded as a theory – the theory that underpins biology. But to be part of science, a hypothesis must be testable by observation or experiment, at least in principle. That is to say, any hypothesis in science must have observable consequences that may either support it or refute it. The philosopher Karl Popper summarised this view as follows "Statements constituting a scientific explanation must be capable of empirical test. The criterion of the scientific status of a theory is its falsifiability, or refutability, or testability".

I discuss in Chapter 4 some evolutionary hypotheses that have withstood attempts to falsify them. Hypotheses made in a supernaturalist framework are not testable, which is why there is so much conflict between different religions – there is no way of resolving differences between them. For example, in the Christian religion alone, there are between 9000 and 33,000 distinct denominations that are recognised, depending on how they are defined. Each of these denominations claims that its interpretation of the Bible is the correct one. Science on the other hand advances cumulatively, step-by-step, reaching broad agreement. Compared with religion, science speaks with one voice, and so is humanity's only universal language.

Science and Religion Compared

That is all I want to say about how science works. Figure 2.10 provides a summary of this topic, taken from the web site of the National Academy of Sciences in the United States. I have discussed points 1–5, but point 6 is also important, because this is where science and religion differ in the modern world. Science tells you how the world works, but it does not tell you how to behave or what to admire – science is morally and aesthetically neutral. The major world religions, on the other hand, do offer instruction in these important areas, but this advice is based on their supernatural interpretations of the world. Because there is no general agreement on the nature and intentions of postulated supernatural agents, it is not surprising that different religions take conflicting moral positions on such things as warfare, the status of women, and sexual behaviour. Science, by comparison, helps you to predict the likely possible results of any given type of action that you are considering.

HOW SCIENCE WORKS
SUMMARY
Source: US National Academy of Sciences

1. Scientists pose, test and revise multiple hypotheses to explain what they observe in the natural world.

2. Scientists use only natural causes to explain natural observations.

3. Science does not prove or conclude; science is always a work in progress.

4. Science is neither democratic nor dogmatic.

5. Scientific claims are subject to peer review and replication.

6. Science is a human endeavour but it cannot make moral or aesthetic decisions.

Fig. 2.10

Some people argue that religions are really just ethical systems aimed at persuading people to behave better, and thus that religious beliefs should be encouraged, but this argument is incorrect. All religions start by postulating explanations about the nature of the physical world and the forces that control it, based on human experience. It is only later that some religions try to justify their beliefs by recommending or enforcing certain types of behaviour.

Religious beliefs do have some major positive effects. For example, many people that run charities helping the poor and disadvantaged are motivated by their religious outlook. It is also obvious that such beliefs have inspired and stimulated many forms of art, especially painting, sculpture, architecture and music. You have only to look up at the Sistine Chapel ceiling in St. Peter's in Rome or listen to the King's College Choir in Cambridge, to realise this. Science has not inspired art to anything like the same extent, but in my personal view, what science is doing is to reveal a Universe whose complexity and beauty surpasses anything imagined by supernaturalists. To appreciate this, look at the breathtaking photographs taken by the Hubble Space Telescope.

On the other hand, science lacks for some, but by no means all, people, the same emotional appeal as religion – it presents a view of human life that is bleak and joyless by comparison, because of the absence of any discernable, overall purpose in the Universe. This view conflicts with the purpose-driven, individual lives that we all lead. This relative lack of appeal is probably the main reason why the majority of people confine their interest in science to its useful applications or dangers, and turn to religion to seek meaning and comfort, especially in times of grief and hardship. Belief in the supernatural provides the possibility of hope in circumstances where a naturalistic approach might provide none. As the poet T.S. Eliot wrote, "Humankind cannot bear very much reality".

The urge to believe in the existence of a personal, all-loving and all-powerful God is very strong in many, but not all, people. The strength of this tendency is shown by the lengths of irrational reasoning that some people will go to in attempting to explain how such a God can permit horrible things to befall innocent people. For example, the argument has been advanced that the Holocaust was permitted by God because he has given humans free will, that is, the ability to make choices between different courses of action. The problem with this argument is that it conflicts with the idea that this God is all-loving, so how can he permit such events, unless he is not all-powerful? This argument also does not explain terrible things that happen, not because of human actions, but because of natural disasters such as earthquakes.

A recent example of this type of thinking was shown by Rowan Williams, the current Archbishop of Canterbury, who was observed to say, when witnessing from close quarters the deliberate destruction of the Twin Towers in New York in 2001, that "God is useless". He later explained that this terrible event had been permitted because God has given us free will. Thus the depth of his need to believe in an all-loving God overrode the simpler explanation of such events provided by the naturalistic viewpoint. On the naturalistic view of the world, such events present no such problem – bad things happen to innocent people because they were unlucky enough to be in the wrong place at the wrong time, while bad behaviour exists because violence, cheating and aggression had survival value during the early evolution of humans, just as it did for other animals.

The Naturalistic Origins of Moral Values

Many people derive their moral values from their religious beliefs. The creationists prominent in the USA reject evolution partly because they fear that acceptance of the evolutionary origin of humans will undermine the basis for morality and lead to social breakdown. They think that if the idea that humans are just another sort of animal becomes widely accepted, it will lead to an increase in violence and disorder. What little evidence there is in the peer-reviewed literature does not tend to support this view, and I will now discuss this evidence as an example of how scientists, in this case social scientists, try to understand the world.

In 2005, a freelance palaeontologist called Gregory Paul published a paper in the electronic peer-reviewed *Journal of Religion and Society,* which is freely available online. This paper compares the incidence of various indicators of the moral state of a society, such as homicide, juvenile and early adult mortality, teenage pregnancy and abortion, and the incidence of sexually transmitted diseases, with the incidence of religious belief and acceptance of human evolution in eighteen prosperous democracies of the world for which extensive data are available. This comparison showed a negative correlation between the acceptance of human evolution and the degree of religious belief. Thus the least religious nation of those surveyed, Japan, shows the highest acceptance of human evolution, while the lowest level of acceptance is found in the most religious developed democracy, the USA.

These correlations support the view of creationists that religious belief tends to lead to the rejection of belief in human evolution, but their further conclusion that therefore the latter leads to a less moral society is contradicted by the data on the incidence of the indicators of low moral standards listed above. All these indicators correlate positively with religious belief, the leader being the USA, which has the highest rates of homicide, early mortality, teenage pregnancy and abortion, and sexually transmitted infection rates in the developed world. The most successful countries by these indicators are the secular democracies, France, Scandanavia and Japan. Some of these correlations are illustrated in Fig. 2.11.

Correlations must be interpreted with great caution because the observation that two phenomena are correlated with each other does not necessarily mean that one causes the other. For example, important causal factors in the high rate of homicide in the USA are likely to be the easy availability of firearms and the wide disparities in wealth between different groups of people. Gregory Paul himself cautions in the Introduction to his paper that "This is not an attempt to present a definitive study that establishes cause versus effect between religiosity, secularism and societal health. It is hoped that these original correlations and results will spark further research and debate on the issue".

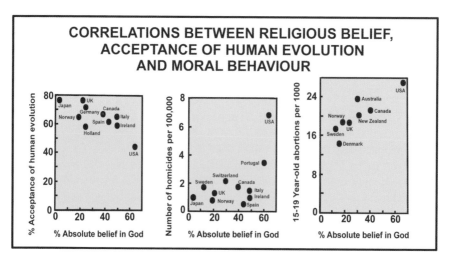

Fig. 2.11

A later paper was published in the same journal in 2006 by an independent sociologist called Gary Jensen. This paper reported a more sophisticated statistical analysis of data from 54 countries collected in the period 1990–1998 about the rates of homicide and different types of religious belief. There are strong positive correlations between homicide rates and the more passionate religious beliefs, such as the belief that the world is a battleground between opposing supernatural forces, often called God and the Devil, and the belief that moral values are so rigid that there is no middle ground between good and evil. People who hold such beliefs are more likely to respond in a violent way to those who oppose them because they think they have supernatural justification for their actions.

However, if we now look at the relation between homicide rates and belief in God but not the Devil, with belief in Heaven but not Hell, and with church attendance, the correlation is no longer positive but negative – such people tend to behave better than the average person because their view of the world is less confrontational and emphasises tolerance. Thus the links between religious belief and homicide rates are clearly much more complex than suggested in the Paul paper but, at the very least, these correlations do not support the commonly-expressed view that a highly religious society is always a morally healthier one. The reasons for this are likely to be many and varied, and deserve much more study by social scientists than they currently receive.

If the naturalistic assumption is correct, moral values must originate from natural sources. An important aim of evolutionary theory is to explain why the vast majority of people have a sense of natural justice, that is, a sense of right and wrong, despite the fact that moral behaviour of that sort is rarely observed in the interactions of non-human animals.

Some evolutionary biologists suggest that morality is a product of natural forces acting through evolution at both the level of individuals and the level of groups of individuals. The basic argument is that those behaviours that increase the probability of survival and reproduction become selected for during evolution. Some of these behaviours are linked to emotional states such as guilt and empathy, so that to us these emotions appear compelling. It is hypothesized that all social animals, from ants to elephants, have modified their behaviour to become less "selfish" because this increases the survival of the group as a whole. On this view, human morality is a natural phenomenon that evolved to increase human co-operation by restricting selfishness.

Individual humans are physically weak and not specialised for running or combat in the way that many other animals are. One of the reasons that, despite these limitations, humans are so successful, is because they co-operate with one another. A simple example is described in the essay *The Biological Basis for Morality* by the biologist, Edward O. Wilson, which is freely available online.

> Imagine a Paleolithic band of five hunters. One considers breaking away from the others to look for an antelope on his own. If successful, he will gain a large quantity of meat and hide - five times as much as if he stays with the band and they are successful. But he knows from experience that his chances of success are very low, much less than the chances of the band of five working together. In addition, whether successful or not, he will suffer animosity from the others for lessening their prospects. By custom the band members stay together and share equitably the animals. So the hunter stays.

We know from experiments with non-human animals that behaviour is partly determined genetically, so if the tendency of humans to co-operate has a genetic component, it follows that genes predisposing people to behave in this way will increase in frequency in the human population. Over thousands of generations, such increases produce those emotions that underlie moral behaviours such as co-operation. In other words, moral feelings are more accurately described as moral instincts. We experience these instincts as conscience, self-respect, shame and outrage. Further discussion of the naturalistic origins of moral behaviour among humans can be found in the book *The Origins of Virtue* by Matt Ridley (see the Further Reading).

A vital by-product of the strong human tendency to co-operate with one another is the development of technology. If you compare the behaviour of humans with that of other animals, it is clear that we have become the dominant species on the planet because we can control our environment by means of technology. Thus the success of the human species today depends upon co-operation, even to the extent of putting the interests of the community above that of the individual. This survival advantage of co-operation would have been especially important when early humans were evolving over several million years on the African savannah. Which brings me to evolution.

Further Reading

1. *How Science Works:* David Goodstein. www.cs.wisc.edu/~dluu/data/papers. A useful discussion about the criteria that distinguish the practice of science.
2. *Religion Explained: The Human Instincts that Fashion Gods, Spirits and Ancestors.* Pascal Boyer. Published by Vintage, 2001. ISBN 0 099 282763. This book discusses the evidence for the intentionality argument for the origin of religious beliefs.
3. *In Gods We Trust: The Evolutionary Landscape of Religion.* Scott Attran. Published by Oxford University Press, 2002. ISBN-13 978-0-19-517803-6. An anthropologist's view of the naturalistic origins of religious beliefs.
4. *Breaking the Spell: Religion as a Natural Phenomenon.* Daniel C. Dennett. Published by Allen Lane 2006. ISBN-13 978-0-713-99789-7. This book argues that religious beliefs originate from the cognitive systems that have evolved in the human brain.
5. *How We Believe; The Search for God in an Age of Science.* Michael Shermer. Published by W.H Freeman and Company, 2000. ISBN 0-7167-4161-x. This book addresses the question as to why people believe in God, as determined from surveys, and examines the validity of the different reasons they give.
6. *The Origins of Virtue.* Matt Ridley. Published by the Penguin Group, 1996. ISBN 0-670-86357-2. This book discusses the origins of co-operation as an evolutionary strategy that led to human society.

Chapter 3
Darwin's Theory of Evolution

Evolution is a vast subject, so what I shall do in the space available is to provide in Chapter 3 a summary of the evolutionary theory proposed by Charles Darwin, and in Chapter 4 some of the principal evidence in support of this theory. For more detailed information on both these topics, you should look at the books on evolution in the Further Reading at the end of each Chapter.

The aim of evolutionary theory is to explain the two most striking features of life on Earth – its astonishing diversity, the fact that there are so many different sorts of living organism, and the exquisite adaptation of each organism to its environment. Let us now look at each of these features in turn.

Biodiversity

Organisms occur in many different environments, from polar seas to tropical forests. Each distinctive environment is called an ecosystem. The biodiversity of a given ecosystem is defined as the total number of different organisms it contains. Figure 3.1 illustrates the diversity of organisms found on a tropical reef at the Great Barrier Reef of Australia.

A major aim of biologists is to characterise and name every different type of organism living in all the ecosystems on the planet today. One reason for this is the belief that the stability of the Earth's ecosystems depends upon their diversity, but that this diversity is being rapidly reduced by human activities that are changing natural habitats at an increasing rate. The loss of tropical rain forests to agriculture and logging is causing particular concern. Many biologists think that the rate of species loss is greater now that at any time in human history, and fear that the global ecosystems on which we depend for our food production may collapse if the diversity is reduced further. One estimate is that 25% of existing species will be lost by the end of the twenty-first century. An additional argument for preserving biodiversity is an aesthetic one – the world would be a less interesting place if there were no tigers, whales or frogs. Several international organisations are dedicated to fighting this reduction in biodiversity.

How many different types of organism are thought to exist? How do we distinguish between different types? These questions are more difficult to answer than

J. Ellis, *How Science Works: Evolution*, DOI 10.1007/978-90-481-3183-9_3,
© Springer Science+Business Media B.V. 2010

Fig. 3.1

you might think. For those organisms that are visible to the naked eye (often called macroscopic organisms), the basic unit of classification is the species. There are several different definitions of a species, but the most widely used is that a species "is a group of actually or potentially interbreeding natural populations that is reproductively isolated from other such groups". Thus all humans are members of the same species, but chimpanzees are a different species. Horses and donkeys are different species because, although they occasionally mate and produce offspring called mules, these offspring are not fertile.

Estimates as to the total numbers of living species that have been described so far range from 1.5 to 1.75 million, and Fig. 3.2 gives an approximate breakdown into the major groups. The reason for this range of values is the varying definitions of species that are used by different biologists. However there is no disagreement that these numbers are much too small, and that many more species remain to be discovered. Estimates of how many more living species remain to be described range from 5 million up to 100 million.

These numbers are large, but the fossil record shows that the number of extinct species is much larger than the number of living species. About half of all the known animal species alive today are insects, and it is thought that many more species of insects exist than have been described, especially in tropical rain forests. The most common insects are beetles. The biologist John Haldane was once asked by a theologian what he had deduced about the nature of the Creator from his study of biology; to which Haldane replied "He must have an inordinate fondness for beetles".

Fig. 3.2

The numbers of different Bacteria and Archaea alive today are thought to be between ten and one hundred times greater than those that have been identified so far. Although some of these micro-organisms exchange pieces of DNA, they do not reproduce sexually in the way that eukaryotic organisms do, so the species definition used for macroscopic organisms is difficult to apply to them. Instead we have to classify prokaryotic species by their biochemical characteristics, such as their DNA sequence. On this basis, a prokaryotic species is defined as cells whose DNA sequence identity is greater than 95%.

Most Bacteria and Archaea cannot be cultivated in the laboratory, so these estimates as to the numbers of different species yet to be studied are based on sequence analyses of the total DNA isolated from terrestrial, aerial and aquatic environments. The most prolific source of different Bacteria and Archaea is the ocean. These organism are prokaryotes – that is, their genetic material is not separated from the cytoplasm by a membrane, as it is in the eukaryotic cells from which animals, plants and fungi are made. Prokaryotic cells were the first type of cell to arise on the Earth, about 4 billion years ago. Prokaryotes were the only form of life for about 80% of the time that life has existed, whereas humans have been around for only about 0.1% of this time. There is a sense in which Bacteria and Archaea still dominate the world. This seems surprising because we cannot see them with the naked eye, but one millilitre of sea water contains between one hundred thousand and one million prokaryotic cells – remember this when you next swim in the sea! Another illustration of this point is that the human body contains about ten times as many bacterial cells as human cells.

Adaptation

One feature of the natural world that much impressed Charles Darwin is how well organisms are adapted by their form and behaviour to survive and reproduce in their natural environment. I will describe here one animal example and one plant example. There is another animal example illustrated in the Frontispiece.

Figure 3.3 shows a woodpecker as an example of a bird adapted to its particular environment.

Fig. 3.3

Darwin was interested in those features of woodpeckers that suit this type of bird to its lifestyle. They have powerful beaks that act as both hammers and chisels, and enable the bird to make holes in trees to use as nests and to find insects living in the wood and bark. The beak contains a long tongue to extract insects from inside the wood. The skull is thick and has a cushion of spongy bone at its base to absorb the shock of hammering. Unlike most birds that have three toes pointing forwards and one backwards, woodpeckers have two toes pointing each way. This arrangement of toes helps the woodpecker to grip rough vertical surfaces. The tail feathers are strong and stiff, providing a brace to help the woodpecker maintain its grip with short legs while making holes in trees. These adaptations all help woodpeckers to survive and reproduce.

Darwin was also interested in adaptations shown by plants, and one group that he studied intensively and described in a book devoted to them, are plants that eat insects. Figure 3.4 is a photograph of such plants that I took in 2008 in Darwin's greenhouse at his home in Kent called Down House.

Fig. 3.4

There are over 600 known species of insectivorous plants and they use different mechanisms to catch and digest insects. Some have sticky leaves and hairs that bind insects tightly, while others have leaves formed into traps that snap shut when an insect lands on them. The species shown in Fig. 3.4 is called *Darlingtonia californica* or the cobra plant, and it uses a more sophisticated trapping device called a pitcher. Pitchers are formed from leaves that have fused at the edges into a tube, and are open near the top where insects such as ants can enter. The top of the pitcher is curved over to prevent rain filling the pitcher. This curved top is speckled with small chlorophyll-free areas, and it is thought that these function to fool the ants into thinking that the clear patches provide exits from the pitcher, so that they exhaust themselves trying to penetrate them. The inner lining of the pitcher contains downward pointing hairs and waxy flakes, so that insects readily fall to the bottom, where they drown in a liquid secreted by the pitcher cells. This liquid contains bacteria and protozoa that breakdown the insects into nutrients that are readily absorbed by the plant cells. These adaptations allow these plants to survive in places where the soil has such low levels of nutrients, especially of nitrogen and phosphorus, that other plants cannot grow. Such places are acid bogs and rocky outcrops. *Darlingtonia californica* is found on bogs and streambanks fed by cold mountain water in Northern California.

Possible Explanations for Biodiversity and Adaptation

The question that biodiversity and adaptation pose is – how can we explain the origin and diversity of all this complexity? Are some or all of this vast number of different organisms related to one another or did they arise separately? Do organisms stay constant in form once they have appeared or do they change over time? By what mechanisms have organisms become so well adapted to their environments? How can we try to answer these questions?

Applying Occam's razor, a good theory should describe one mechanism to account for the two most distinctive aspects of life: the very large number of different organisms that exist today, and the adaptation that each shows to the environment in which it thrives. Three general hypotheses have been proposed at various times to explain the diversity of life (Fig. 3.5).

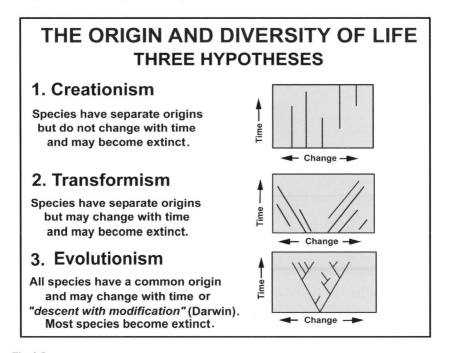

Fig. 3.5

Creationism – this is separate creationism, not to be confused with religious creationism, also known as creation science or intelligent design. Remember that we are working within a naturalistic framework. This hypothesis supposes that each species had a separate origin by some unspecified natural mechanism and does not change with time, but may become extinct.

Transformism supposes that species have separate origins, possibly at different times, but can change with time, and may become extinct.

Evolutionism supposes that all species have a common origin, can change with time and may become extinct.

Each red line in Fig. 3.5 represents a single species. Vertical red lines indicate that species do not change with time between generations, while red lines sloping either to the left or to the right indicate that species do change with time between generations. If you compare these three hypotheses in terms of Occam's razor, you see that evolution is simpler because it supposes only one origin of life, while the other two suppose as many origins as there are species. None of these hypotheses however, explains either how life originated or how the diversity arose. Explaining the origin of the first living cells is one of the most challenging unsolved problems in biology, but explaining the origin of life's diversity is one of the intellectual triumphs of science. This explanation is called the theory of evolution by natural selection, proposed by Charles Darwin.

Prior to Darwin, most people thought that each species was created separately and persisted unchanged. Religious people thought in addition, that the act of creation was performed by a supernatural entity of some sort, but remember, science works within a naturalistic framework and looks for natural causes of natural events. So the problem is to find a testable natural explanation of the huge diversity found in the living world. The explanation that is accepted by biologists today was suggested by Charles Darwin in a book, the full title of which is *On the Origin of Species by Means of Natural Selection.* The first edition was published in 1859 and the sixth edition in 1872. The 200th anniversary of Darwin's birth and the 150th anniversary of the first publication of his famous book occurred in 2009, and these anniversaries were celebrated around the world. The publication of *On the Origin of Species* marks a turning point in the history of science – the point where nature study became the unified discipline that we now call biology.

The Early Life of Charles Darwin

Charles Darwin was an English naturalist born into a wealthy family living in Shrewsbury. When young, he had a passion for all things outdoors, and collected birds' eggs and minerals. At the age of nine he entered Shrewsbury Grammar School where he was taught Greek and Latin, which he hated and learnt very badly, but received no tuition in either science or sports. The headmaster described him as "a very ordinary boy, rather below the common standard in intellect" so his father withdrew him from school at the age of 16. His father was a popular and successful doctor, and wanted Charles to follow in his footsteps, so he sent him to Edinburgh University where Charles' brother was studying medicine. Unlike today, you did not need any exam passes or any qualifications at all to enter university – all you needed was the money to pay the tuition fees! You also had to be male – girls were not allowed to attend university at this time.

To his father's dismay, but fortunately for us, this plan failed. Charles was appalled by the sufferings of patients undergoing surgery – anaesthetics were not used in medicine until the end of the 1840s. So he left the medical school in Edinburgh, and went to Cambridge in 1828 to study for becoming an Anglican parson, which was a common fate for wealthy young gentlemen in his day. While at Cambridge he became friendly with two professors – Adam Sedgwick, the professor of geology, and John Henslow, the professor of botany. He did well in his final examinations in 1831, but then went on a summer field trip with Sedgwick to North Wales, to learn geology first hand from an expert. He planned to use this geological expertise for a trip he was thinking of making to the Canary Islands in order to study the natural history of tropical regions, before becoming a country parson. But on returning from this field trip, Darwin found a letter from Henslow proposing him as a suitable gentleman naturalist and companion to Captain Robert Fitzroy on the HMS *Beagle*, an Admiralty ship that was about to leave on an expedition to chart the coastline of South America. This was perhaps the most important letter in the history of science, because it resulted in Darwin abandoning his father's hope that he become a country parson, and instead to become one of the most influential people in the history of human thought.

HMS *Beagle* left Plymouth in December 1831 for what was planned to be a two-year voyage, but returned nearly five years later in 1836, after sailing around the world. The ship visited not only both the East and West coasts of South America, but also the Galapagos Islands, Tahiti, New Zealand, Australia, Mauritus, Capetown and various other islands. Darwin made many trips inland, observing and collecting plant and animal specimens new to science that he sent back to Henslow in England, as well as recording the geology of these regions. Tropical rain forests especially delighted him, as they were rich in plants and insects previously unknown, as they still are today. He witnessed earthquakes, found fossils of large extinct mammals that looked curiously like some much smaller mammals around today, such as armadillos, and discovered the shells of marine organisms at the top of mountains.

While at Cambridge, Darwin was required to read books by the Anglican philosopher William Paley. One such book was entitled *Natural Theology*, first published in 1802. In this book, Paley argued that observations of the living world strongly indicated that it had been deliberately created by an intelligent supernatural being. His basic argument stems from the obvious fact that living organisms are much more complicated than non-living objects like rocks, and moreover their complexity enables them to survive and reproduce their kind in the particular environments in which they live. Today we express this by saying that organisms are highly adapted to their environments. Paley then used a metaphor to explain his reasoning – the metaphor of a watchmaker – and his argument is referred to as the Argument from Design.

Paley pointed out that if we encounter a rock during a country walk, its nature raises no particularly puzzling questions. But if we found a watch, and had not seen one before, we would observe that, compared to the rock, the watch is much more complicated, and moreover that this complexity is directed to serving a specific function – that of indicating the time. Paley argued that the only explanation must be that an intelligent entity had designed and built such a complex purposeful structure. He then extended this conclusion to explain the exquisite adaptations of all animals and plants to their environments in terms of their creation by God.

In his *Autobiography* Darwin explained that he enjoyed reading Paley's books during his undergraduate days at Cambridge and was "charmed and convinced" by Paley's arguments. But his experiences during the *Beagle* voyage led Darwin to question Paley's supernatural argument from design, and he started to wonder what natural biological and geological events could more simply explain the diversity, adaptations and geographical distributions of so many different organisms, both now and in the past. On his arrival in England, he decided to devote the rest of his life trying to understand why the living world looks and behaves as it does. His sister Caroline recorded at this time that Charles had gained "an interest for the rest of his life", while Darwin later recorded in his *Autobiography* that this voyage "determined my whole career".

Darwin did not invent either the idea of evolution or the idea that he called natural selection. The idea that species may evolve can be traced back as far as the sixth century BC to the Greek philosopher Anaximander, who proposed that the first men were generated in the form of fish. This view was not associated with any religious belief, so it represents the first known example of evolutionary thinking in a naturalistic framework. What Darwin did in his book was to propose a particular mechanism, and to amass a large amount of evidence in support of that mechanism – the process he called natural selection. Unknown to Darwin, other people had suggested this sort of mechanism before him, but unlike Darwin, they had failed to present convincing evidence that supported this idea. So Darwin's real contribution was to be the first person to bring together the previous ideas of evolution and natural selection into a single theory, and to provide overwhelming evidence in their favour. The sheer quantity and quality of the empirical evidence that Darwin amassed in favour of the idea of natural selection is impressive even today. Richards Dawkins brilliantly captured the replacement of Paley's idea that God is the watchmaker of the living world with Darwin's idea that natural selection is the watchmaker in the title of his book *The Blind Watchmaker,* first published in 1986. Applying Occam's razor, the design of living organisms is more simply explained by natural selection than by the actions of an intelligent supernatural being because natural selection involve fewer assumptions and these assumptions are testable.

Thus the major impact of Darwin's ideas has been to undermine natural theology as espoused by Paley. In his *Autobiography*, Darwin says:

> The old argument from design, as given by Paley, which formerly seemed to me so conclusive, fails, now that the law of natural selection has been discovered. There seems to be no more design in the variability of organic beings and in the action of natural selection, than in the course which the wind blows.

Today, the main-stream Christian religions accept that evolution has occurred by natural selection, but assert that this is the way their God has created the living world. This is wise of them, because their cause would be greatly weakened if they did not accept evolution in the face of all the evidence in support of it.

Evolution by Natural Selection

The basic idea that Darwin, and others before him, presented is very simple. Because organisms compete with one another for resources, individuals that are better adapted to their particular circumstances will leave more offspring, and therefore it follows that those better adaptations that are inherited will increase in frequency from one generation to the next. The argument is spelt out in Fig. 3.6 in the form of the four postulates made by Darwin in his book *On the Origin of Species*. The most important feature of these postulates is that they are testable – they contain no hidden assumptions or require anything to be accepted uncritically. Most of Darwin's book is concerned with presenting a large range of direct observations taken from nature that support these postulates. Since his time, many more direct observations from both nature and experiments have reinforced this support.

> # THE DARWINIAN HYPOTHESIS
> ## EVOLUTION BY NATURAL SELECTION
>
> ### Darwin`s Four Testable Postulates
> **(Source: *On the Origin of Species* 1859)**
> 1. Individuals in a population of a given species are variable.
> 2. Some of this variation is heritable.
> 3. In every generation, some individuals are more successful at surviving and reproducing than others.
> 4. Survival and reproduction are not random but depend upon individual heritable variation i.e. `survival of the fittest`.
>
> ### Modern Definitions
>
> **EVOLUTION** - Change in the form and behaviour of organisms between generations
>
> **NATURAL SELECTION** - Change in the genetic composition of populations caused by differences in survival and reproduction

Fig. 3.6

In Fig. 3.6, you will also find definitions of the terms "evolution" and "natural selection". These are modern definitions – the term "genetic" was invented after Darwin's time. Darwin knew nothing about genes or the mechanism of heredity. His library does not contain any papers by the father of genetics, Gregor Mendel, who discovered the particulate nature of inheritance in 1865, just after the first edition of Darwin's book appeared, but who published his work in a little-known journal. Five further editions of Darwin's book appeared, the last in 1872, but none of them refer to the work of Mendel. Unlike Darwin, we know today that inherited adaptations are encoded in genes, which are made of DNA. It follows that the genetic composition of the population changes with time – this is what we mean by the essence of evolution today.

I shall now briefly discuss each postulate in turn, and then summarise studies over the last thirty five years of finches in the Galapagos Islands as an example of a natural population where these postulates have been found to hold.

Postulate 1: Individuals in a population of a given species are variable.

We are all aware that the individuals in any group of people differ from one another in many respects such as height, hair, eye and skin colour, facial appearance, athletic ability, temperament, personality and intelligence. Individual variation is in fact a universal feature of all species, and occurs at all levels from morphology to DNA sequence.

Postulate 2: Some of this variation is heritable.

Not all the differences between individuals are inherited, but it is the variations in DNA sequence between different individuals that produce all those differences that are inherited. Differences that are not inherited can be caused by different behaviours. For example, the different eye colours of a blue-eyed athlete and a brown-eyed sedentary person are due to inherited variations in their DNA, but their different muscle development reflects their attitude to exercise and is not inherited. Thus a blacksmith who develops strong arm muscles will not pass this character directly to his children, but he may encourage them to follow his trade, and his children will then develop muscular arms.

Postulate 3: In every generation some individuals are more successful at surviving and reproducing than others.

Your knowledge of people and their history will confirm the correctness of this postulate for humans, and there is now abundant evidence from observations of other species that is applies to them as well. The average Atlantic female cod fish produces about two million eggs in each breeding season. About 99% of the hatchlings from these eggs are eaten in their first month of life, while 90% of the remainder do not survive beyond their first year of life. On average, each female cod produce only two offspring that survive to reproduce. If she produced less than two such offspring, the population would die out, because it takes two fish to produce eggs. If she produced more than two on average, the population would rise to infinity or until the food supply was exhausted. This production of more individuals than can survive to reproduce is a universal feature of all species because the world does not contain enough resources to support all the new individuals that are born. In his book, Darwin presents a similar calculation for elephants, which reached the same conclusion.

Postulate 4: Survival and reproduction are not random but depend upon individual variation. This situation is summarised by the phrase "survival of the fittest".

This phrase was proposed by the English philosopher, Herbert Spencer, in 1864 after he had read *On The Origin of Species,* and was subsequently used by Darwin. The word "fittest" here does not mean the individual who is athletically superior, as it does in normal language. Instead it means that people who are better adapted to their environment, for whatever reason, have a greater chance of leaving more children that those who are less well adapted. We say such people are "fitter" in the evolutionary sense because there are more copies of their genes in the next generation. Fitness in the Darwinian sense is defined quantitatively as the mean number of offspring left by an individual relative to the number of offspring left by an average member of the population.

Unfortunately, the phrase "survival of the fittest" was used in the twentieth century for political purposes to justify the mistreatment and murder of those deemed "unfit", a practice that would have appalled Darwin. A cousin of Darwin called Francis Galton proposed in 1883 the idea of eugenics – that human breeding should be controlled to allow the reduction of unfavourable inherited traits and the increase of favourable ones. Eugenics was one of those ideas that sounds fine in principle but proved disastrous in practice because it was used to justify violence. Regimes such as that in Nazi Germany argued that because people are clearly not genetically equal, they should not be treated equally under the law – that it was permissible to kill people with characteristics they did not like because this is how evolution worked. Compulsory sterilization of thousands of people with genetic defects was practised in the United States, Germany and Scandanavia during the first part of twentieth century. This practice is now regarded as a crime against humanity by the International Criminal Court.

Using evolutionary theory to justify political actions is an example of a type of unjustified reasoning known as the "is-ought problem". The Scottish philosopher David Hume (1711–1776) was a leading Enlightenment figure who pointed out that there is no justification to decide how the world *ought* to be from how the world *is*. That the appearance of humans is the end result of a process of natural selection that involves massive suffering and violence does not justify humans treating fellow creatures in a similar fashion. On the contrary, the art of civilization consists of humanity striving to rise above its biological past. However, the basic idea of eugenics has not gone away. Developments in genetic technology in the twenty-first century are raising the question as to what extent we should take evolution into our own hands by altering human genomes directly.

Figure 3.7 illustrates the basic principle of natural selection by describing the result of mutations producing changes in the coat colour of lions.

Lions are carnivores and their survival depends upon their ability to catch and eat other animals. Hunting is not easy because the prey animals will survive and evolve only if they are successful in avoiding the lions. Thus there is an evolutionary arms race between hunter and hunted. Coat colour is important to lions because it provides the camouflage that enable them to get closer to the prey before the latter detect their presence. Changes in coat colour that lower the efficiency of camouflage will reduce the ability of the lions to survive long enough to reproduce, but any mutation that increases the efficiency of camouflage will have the opposite effect. In time, the mutant variant lions are so successful that they replace the descendants of their ancestor with the original coat colour – both the original ancestors and the intermediate variants on the way to the successful variants that occur today become extinct. This replacement by more successful descendants explains the lack of intermediate

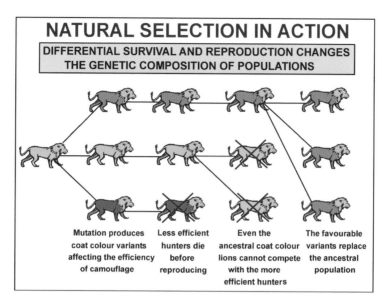

Fig. 3.7

forms in current populations, and indicates that extinction is an inherent part of the process of evolution. The same argument applies to the lack of intermediate species in the fossil record. Thus the genetic composition changes with the generations, as predicted by evolutionary theory, but the vast majority of species that have been created by evolution are now extinct.

Direct Observation of Natural Selection in Finches

In 1973 Rosemary and Peter Grant started a study of natural selection among the finch populations that inhabit the Galapagos Islands in the central Pacific Ocean, a study that continues to this day. These birds invaded the islands from America some two million years ago, and today are divided into 13 distinct species. Comparisons of their DNA sequences show that they are all closely related. They are all similar in body size and colour, but differ in the size and shape of their beaks because they differ in what they eat. One small island called Daphne Major makes a good natural laboratory because it is small and isolated, and contains only around 1200 birds, all of which have been tagged by the Grants and their colleagues, and their beak depths and other bodily features measured. The depth of an individual's beak is important because it is correlated with the size of the seeds that the bird can crack – birds with bigger beaks eat bigger seeds.

To test Postulate 1, every finch was measured for the size of its beak. Figure 3.8a shows a graph of beak depth plotted against the number of finches, and you can see that this character is indeed variable. Is this variation due to inherited differences or to environmental effects or to both? The researchers measured the average beak depth of families of birds after they had reached adult size, and compared it to the average beak depth of their mother and father. Figure 3.8b shows that parents with shallow beaks tend to have young with shallow beaks, and parents with deep beaks tend to have young with deep beaks. Thus there is a large genetic component to the determination of beak depth, supporting Postulate 2.

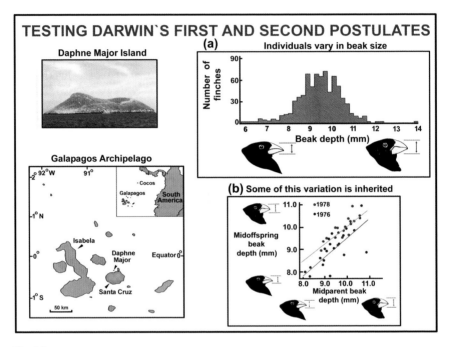

Fig. 3.8

The Galapagos Islands have a variable climate, and experience both droughts and periods of excessive rainfall that affect the food supply of the finches. In 1977 the rainfall was only about 20% of the normal rainfall so that the plants, whose seeds are the main food of the birds, were much sparser than usual. As a result, 84% of the birds died, as shown by the upper graph in Fig. 3.9a. It is the few survivors that allowed the population to recover to normal levels after the drought had ended, supporting Darwin's third postulate.

To test Postulate 4, we now need to ask whether those finches that survived the drought were a random selection of all the finches or whether they were they a sub-set, selected because they possessed some advantage. Small, soft seeds are preferred to large, hard seeds because the finches find them easier to crack. But the types of seeds available as food changed as a result of the drought. The lower graph in Fig. 3.9a shows that the proportion of large, hard seeds increased and that of small, soft seeds declined. But only birds with deep beaks can crack the larger seeds. So as the drought progressed, more of the birds with deep beaks survived than birds with less deep beaks, as shown in Fig. 3.9b, where the arrows on the horizontal axes indicate the average values for beak depth. The birds with deeper beaks also tend to be bigger in body size, so tend to win fights with smaller birds over the supply of seeds. Comparison of beak depth between finches hatched before the drought with those hatched after the drought revealed that the average beak size was increased after the drought. Thus changes in the environment cause changes in the genetic composition of the population. These observations in the field confirm the prediction that natural selection causes populations to change their genetic composition – *natural selection causes evolution*.

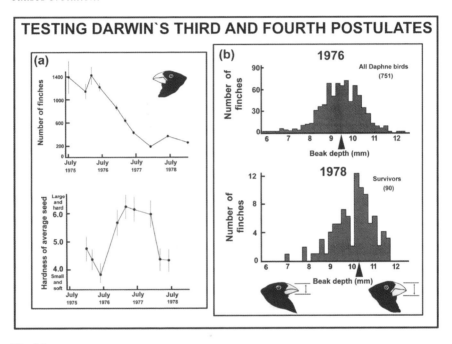

Fig. 3.9

Historical Context of the Idea of Natural Selection

In the sixth edition of *On the Origin of Species* (1872), Darwin explained that he thought of the idea of natural selection in 1838, after reading a book by Thomas Malthus entitled *An Essay on the Principle of Population*. In this book, Malthus pointed out that human populations, if unchecked by disease or conflict, tend to increase at geometric rate (i.e. 1,2,4,8,16 etc), but that food supply increases at a arithmetic rate (i.e. 1,2,3,4 etc). The result is that the size of populations is restrained by the food supply and many people end up in poverty. Since Malthus' time, food production has in fact kept up with the huge growth in the human population. The reason that some people starve is not that too little food is produced, but that they cannot afford to buy it. This situation may well change in the future, as a result of the effects of climate change on agriculture.

At least two other people had previously published the idea of natural selection before Darwin, such as William Wells in 1818 and Patrick Matthew in 1831. Matthew even used the term "the natural law of selection" but his work was not widely known, and Darwin did not come across it until after the first edition of *On the Origin of Species* was published in 1859. In the Historical Sketch that Darwin added to the third edition of his book, Darwin lists Matthew among thirty-four authors who had previously published similar ideas. By 1844, Darwin had completed the first draft of his ideas on natural selection in the form of an essay of 239 pages, but fearful of the controversy these ideas might arouse at such a turbulent time in European history, he placed it in the hall cupboard at Down House, with a note saying that the contents should be published only in the event of his death. When visiting Down House in 2008, I came across a replica of this package, placed in the hall cupboard by English Heritage, who have restored the house and garden to its state in Darwin's time.

Darwin was a prolific letter writer, and corresponded with many people about biological matters. One of these people was Alfred Russel Wallace, a prominent naturalist who published a popular book about the plants and animals of the Malay Archipelago. Wallace was the leading expert on the distribution of animal species in the nineteenth century, and has been called "the father of biogeography". Like Darwin, he also was greatly influenced by reading the essay on population growth by Malthus. In 1858, Darwin was shocked to receive an essay from Wallace, outlining ideas about natural selection closely similar to his own. Realising that he might be scooped, Darwin sent the essay, together with one of his own, to scientific friends in London, who arranged that both papers were read in the absence of their authors at a meeting of the Linnean Society on July 1st, and subsequently published. This unexpected development spurred Darwin to write *On the Origin of Species,* which appeared the following year. Wallace later acknowledged that Darwin had the idea of natural selection before him, and had amassed much more evidence in support. Wallace famously called Darwin "the Newton of Natural History" because he had unified biology with a universally applicable idea, just as Newton unified the physics of his time with his theory of gravity.

This story tells us something about what motivates scientists. As well as the urge to understand the world, each scientist feels the need to be thought well of by fellow scientists, and to achieve this they are keen to publish new ideas and discoveries as soon as possible in order to claim priority. So science is not the disinterested pursuit of truth that it is often portrayed to be – scientists are human, and have egos like everybody else! The same motivation is, of course, present in enthusiasts in other fields, such as literature, music and the theatre.

By 1870, the fact that evolution had occurred was generally accepted among biologists, but the suggested mechanism of natural selection was not generally accepted until the 1930s, some 60 years later. The reason was that Darwin was unable to explain how the variation among individuals was generated, or how this variation was passed onto the next generation – he had no theory of heredity. It is one of the great "Ifs" of history to wonders how the subject would have developed if Darwin had read and grasped the significance of the work of Gregor Mendel. To be fair, no other biologist grasped the significance of Mendel's experiments on heredity in pea plants either, until his work was rediscovered and extended by other biologists at the start of the twentieth century.

Mendel's discovery that heredity in pea plants is particulate in nature was extended and confirmed in the period 1900–1940 by many more experiments with animals and plants, especially the fruit fly *Drosophila*. But it was only in the latter decade of this period that Mendelian genetics was found to be consistent with the idea of natural selection. Three biologists in particular showed by mathematically based studies that the behaviour of genes in populations could be accounted for quantitatively by natural selection – these were Ronald Fisher, John Haldane and Sewall Wright. The fact that these three scientists worked independently of one another strengthened their case – you will recall my earlier point that it is an essential feature of how science works that new claims should be confirmed by independent researchers. The classical work of these three people demonstrated that natural selection could work with the kinds of variation observed in nature by applying the laws of Mendelian genetics. Darwin's theory of evolution by natural selection has rested upon firm genetical foundations since that time. This achievement of the human intellect is often called "The Modern Synthesis", after the title of a book published in 1942 by the biologist Julian Huxley, a direct descendant of Thomas Henry Huxley, one of Darwin's strongest supporters after his return from the voyage in HMS *Beagle*.

Common Misconceptions About Natural Selection

It is important to grasp two particular aspects of natural selection because these are often misunderstood.

Firstly, natural selection acts directly on individuals, but it is only populations that evolve, not individuals. It is often thought incorrectly that evolution applies to individual organisms, perhaps because changes in individuals produced during their development are confused with evolutionary changes. This is another area of science where precise definitions are vital! To understand how populations evolve requires some grasp of statistics, and some people find it difficult to think in statistical terms.

Secondly, natural selection responds to pressures produced by the immediate environment. So natural selection is not random but neither is it a directed process, because it has no foresight, no overall progress in a particular direction. It cannot predict the future – how could it? *Any* inherited variation that promotes an individual leaving more offspring than its competitors in the present environment will be selected for. If the environment changes, some of these variations may then be selected against, while other variations may be favoured. For example, in 1983 the rainfall in the Galapagos Islands was 57 times greater than in 1977, the year of the drought. This excessive rain resulted in a plentiful supply of small, soft seeds which the finches harvest more efficiently than large, hard seeds. So after wet years, the smaller finches with shallow beaks reproduce more rapidly than larger birds with deeper beaks – exactly the opposite of what happens in drought years. Evolution by natural selection is a highly dynamic process, but it is always at least one generation behind changes in the environment.

Adaptation by natural selection often involves an increase in complexity because in many environments, complexity increases fitness. In Victorian times many people thought that evolution is *always* progressive – after all, it had produced human beings such as themselves who are clearly superior to all other forms of life! But this is another misunderstanding of how evolution works.

Complexity evolves only because it improves the fitness of the next generation in the current environment. But if the environment changes, complexity may become a disadvantage, and so natural selection can lead to a reduction in complexity. For example, many parasites adapt to their environment by losing systems they no longer need because their host can replace them. The tapeworm has no gut because it no longer needs one in its present environment, living inside the digestive systems of its host, but it evolved from a free-living worm that did have a gut. Some fish that live inside caves have eyes that cannot see, because they have evolved from fish that had functional eyes. Snakes have no functional legs but evolved from animals that did, as shown by the presence of tiny hind limbs in some species. Many people, including some biologists, find it difficult to abandon a progressive view of evolution, and persist in using terms like "higher" and "lower". Thus a moss is said to be a "lower plant" and a daisy a "higher plant". But every organism alive today is as much evolved as any other organism. It is just that some are more complex than others.

Many people find this directionless aspect of the evolutionary process disagreeable, because it offends our innate sense that human beings like us are so wonderful that the process that produced us must be purposeful. This is really an example of human arrogance and intentionality – we are conceited enough to feel we must be special. Evolutionary theory, in contrast, says that we are just another animal, produced by the same process that has produced all the other animals, as well as the plants, fungi and micro-organisms.

Another aspect of evolutionary theory that disturbs many people is the sheer amount of pain and suffering that the mechanism of natural selection has produced in the animal kingdom over hundreds of millions of years. Natural selection works by eliminating the less fit, and the agents for this include disease, starvation and predation. It is for this reason that some biologists, such as David Attenborough, have stated that they are not able to accept the idea that the living world was created by an omnipotent, beneficient creator. Scientists who do accept this idea often respond to this problem by saying that the actions of their creator cannot be understood by mere humans.

Genetic Drift

Darwin thought that natural selection was the main cause of evolution but since his time another mechanism has been discovered, called genetic drift. This process is defined as the change in gene frequency between generations caused by random sampling effects. The word "drift" is unfortunate because it could be interpreted to imply direction, but the effect is purely random, unlike natural selection, which is highly non-random.

Suppose two men are in the forest, collecting wood. One man has a set of genes conferring high evolutionary fitness i.e. he is capable in principle of fathering six children, but has fathered only one so far. The other man is less fit, because his genes do not result in him being able to pass so many copies of his genes to the next generation and he has no children. Suddenly a storm blows up and a tree falls on the first man, killing him. The second man goes on to father two children, so the genetic composition of the next generation is different from what it would have been if the first man had survived to father five more children. A purely random event has changed the genetic composition of the population produced by these two men.

Biologists continue to argue about the relative effects of natural selection and genetic drift in evolution, but what is agreed is that genetic drift will be more important in small populations than in large populations. A human example of genetic drift involves a small group of people that crossed the Bering Strait between America and Russia about 10,000 years ago, at the end of the last Ice Age. This group gave rise to the Native Americans that live in the USA and South America today. The observation that these Native Americans almost totally lack the gene for the protein determining blood group B suggests that this founding group was very small in number because about 16% of the entire world's population of humans possesses this blood group. This type of genetic drift is called the founder effect for obvious reasons, and is commonly seen in island species that are descended from a group of organisms too small in number to contain a representative sample of all the genes present in that species on the mainland.

Further Reading

1. *Introducing Darwin.* Jonathan Miller and Borin van Loon. Published by Icon Books 1982. ISBN 1 84046 715 0. An account in cartoon form of the impact of Darwin's work on biology.
2. www.darwin-online.org.uk This website contains all the publications of Charles Darwin.
3. www.darwinproject.ac.uk This website contains all the correspondence of Charles Darwin.
4. http://evolution.berkeley.edu An Introduction to evolutionary theory from the University of California at Berkeley.
5. *Evolution:* Nicholas H. Barton, Derek E.G. Briggs, Jonathan A. Eisen, David B. Goldstein and Nipam H. Patel. Published by Cold Spring Harbor Laboratory Press, 2007. ISBN 978-0-87969-684-9.
6. *Evolutionary Analysis:* Scott Freeman and Jon C. Herron. Published by Pearson Education, Inc. 2007. 4th edition. ISBN 0-13-239789-7.
7. *Evolution:* Mark Ridley. Published by Blackwell Publishing, 2004, 3rd edition. ISBN 1-4051-0345-0.
8. *Understanding Evolution: History, Theory, Evidence and Implications.* Geoff Price, 2006. www.rationalrevolution.net/articles/ – scroll down to 'Understanding Evolution: History, Theory, Evidence, Implications'. This is a comprehensive account of the historical relations between science and religion in the context of evolution.

9. www.talkorigins.org/ A collection of essays and articles that provides mainstream scientific responses to questions raised in the debate about creationism and evolution as alternative explanations of the living world.
10. *Evolution:* Brian and Deborah Charlesworth. Published by Oxford University Press, 2003 in the Very Short Introduction Series. ISBN 978-0-19-280251-4. A succinct and pocketable summary of what is known about evolution.

Chapter 4
The Evidence for Evolution

Now I want to turn to the evidence for evolution. If you ask the average person-in-the-street about this, they will probably mention the fossil record – the remains of types of organism no longer living on the Earth. But this is incorrect – the fossil record is consistent with both separate creationism and with transformism, which, you will recall, both propose that species had separate, natural origins but can become extinct. Darwin did not use the fossil record as one of his main lines of evidence to support the idea of evolution because he thought that there were not enough fossils known in his time. He devoted a whole chapter in *On the Origin of Species* to this problem, stressing the absence of numerous transitional forms between fossil species and species alive today that his theory predicted should occur. He suggested that this absence could be explained if the fraction of organisms that end up as fossils is extremely small and dependent on particular geological events that themselves vary with time. He lists at the end of the chapter the names of nine eminent scientists that specialised in studying fossils in his day but who rejected the idea that species had changed over geological time.

What the fossil record shows is that in the past there were organisms that are not around today, but some of the fossils discovered since Darwin's time are of organisms that show features intermediate between those of the major groups of organism alive today. For example, some fossils are of animals that had feathers and wings, like birds do today, and teeth and bony tails, like reptiles do today. Such transitional features are expected if evolutionary theory is correct, but are not strictly ruled out if either separate creationism or transformism is correct. So, while the fossil record is consistent with evolution, of itself it does not exclude other logically possible naturalistic explanations. So what is the evidence for evolution?

Figure 4.1 lists seven of the main lines of evidence for evolution.

The first line of evidence concerns similarities between organisms that you would not expect if they had independent origins. These similarities are found at all levels, from the molecular to the anatomical; I will show you examples from two levels. These similarities are often referred to as "homologies", but there is a possible

J. Ellis, *How Science Works: Evolution*, DOI 10.1007/978-90-481-3183-9_4,
© Springer Science+Business Media B.V. 2010

SOME EVIDENCE FOR EVOLUTION

1. SIMILARITIES: All organisms show similarities at many levels that would not be expected if they had independent origins

2. DIRECT OBSERVATION: Changes in gene composition between generations can be seen in rapidly reproducing organisms

3. TRANSITIONAL FOSSILS: Fossils of forms intermediate between one type of organism and another type of organism living today have been found

4. LOGICAL INFERENCE: Evolution is the inevitable consequence of natural selection acting upon the effects of mutations in DNA

5. HIERARCHICAL CLASSIFICATION: Organisms can be classified into groups within groups, as expected if they are all related by descent

6. BIOGEOGRAPHY: Location provides a better index of biological similarity than does similarity of climate because geography reflects descent from common ancestors

7. VESTIGIAL ORGANS AND FUNCTIONLESS GENES: Features that become useless are slowly lost, becoming vestiges of ancestral forms

Fig. 4.1

source of confusion with this term because its meaning has changed over time. "Homology" was coined by a contemporary, and bitter rival, of Charles Darwin called Richard Owen. Owen was an anatomist, so he defined the term in 1843 to mean "the same organ in different animals under every variety of form and function". Unfortunately, some biologists today use the term "homology" to mean "evolutionarily related", so to cite homology in the latter sense as evidence for evolution is a circular argument.

Then we have direct observation, as the second line of evidence. The rate of evolution depends upon the rate of reproduction, and this is much too slow in the case of animals like ourselves to be seen directly, but in the case of very rapidly reproducing organisms like viruses, bacteria and some insects, changes in the genetic composition can be seen between generations. Thirdly, if one type of organism has changed into another type of organism over long periods of time, there should be transitional forms sharing characters from both types in the fossil record. I will show you some examples that bridge the gap between reptiles and birds, and the gap between fish and amphibians.

Then we have logical inference – the fourth line of evidence. Evolution depends upon two processes – the mutation of DNA, and natural selection operating on the effects of these mutations on the properties of the organism. Both these processes are directly observable today, so it logically follows that evolution must have happened if these processes occurred in the past. In fact, if we did not have strong evidence that evolution had happened, and is still happening, we would have a problem to explain why evolution has not occurred!

The fifth line of evidence is the fact that organisms can be classified in an hierarchical fashion in a nested pattern of groups within groups, as predicted by evolutionary theory. The sixth line of evidence is the one favoured by Darwin, and comes from biogeography, the study of the distribution of organisms across the world. The observation that the most closely related species are found close together geographically, regardless of their habitat or their specific adaptations, is explicable in terms of evolutionary theory. Finally, the existence of vestigial organs and functionless genes is consistent with their origin from earlier organs and genes that were functional, but hard to understand if species were separately created,

Similarities at the Molecular Level

The Unity of Biochemistry

Figure 4.2 shows some similarities at the molecular level. There is a sense in which the huge diversity of living organisms that is apparent to the naked eye, and which delights us all, is an illusion, because this diversity becomes much smaller when we compare organisms at the molecular level. This fact is summarised by the phrase "The Unity of Biochemistry". So the basic metabolic pathways, the energy transduction mechanisms, the signalling systems, and the operations of replication, transcription and translation are very similar in all organisms, no matter how different these organisms look to the eye. A few examples from many that can be given

EVIDENCE FOR EVOLUTION
SIMILARITIES (or HOMOLOGIES)
Organisms show similarities at many levels that would not be expected if they had independent origins
Similarities at the molecular level

THE UNITY OF BIOCHEMISTRY
The huge diversity of organisms obvious to the naked eye conceals the surprising fact that all the basic biochemical processes are very similar in all organisms

EXAMPLES: 1. All cells use DNA as the genetic material.
2. The basic machineries of replication, transcription, and translation are the same in all organisms.
3. The genetic code is almost universal.
4. All cells use ATP to drive metabolic processes.
5. All proteins are made from the same set of twenty L-amino acids.

Fig. 4.2

are listed in Fig. 4.2. The continuing success of the discipline of molecular biology is only possible because of this basic similarity between all organisms. It is the unity of biochemistry that enables genetic engineers to take a gene from one type of organism and get it to work inside a quite different organism. For example, the production of human insulin by both bacteria and yeast has been achieved by isolating the genes for this vital hormone from human cells and inserting them into bacterial and yeast cells. These methods of production have now replaced the original method of isolating insulin from the pancreas of pigs and cows for the treatment of human diabetes.

The most detailed information about the diversity and relatedness of organisms available today is provided by determining the base sequence of the DNA found in different organisms. The total DNA sequence of each species is called its genome, and it is the collection of genes in this genome that contains the information to build and operate that species. Due to advances in sequencing technology, it is now possible to determine the total DNA sequence of say, a new type of bacterium, in 24 hours. In recent years, increasing number of total genome sequences have been determined, mostly for bacteria, but including some plants and animals, including humans, chimpanzees, dogs, cats, fish, worms, flies, rice and maize. These sequences show directly that, for example, we share almost all our genes with chimpanzees, and thus chimpanzees are regarded as our closest relatives. But we also share many of our genes with bacteria and plants – all organisms are genetically related. None of these similarities is predicted by hypotheses that organisms have been separately created.

The Principles of Molecular Biology

To help you understand how the DNA sequence in the genome specifies the entire organism, Fig. 4.3 reminds you of the basic principles of molecular biology.

The modern view about the nature of life is that organisms are self-assembling chemical machines programmed by their genes. The term "self-assembling" is used because organisms grow, develop and reproduce by taking in chemicals from the environment, and converting them into the huge range of other chemicals necessary to build cells. The sum of all these processes is called metabolism. All the information for metabolism is contained within the collection of genes that each organism inherits from its parents. This information is internal to the organism and requires no external directing agency for it to operate – in other words, organisms assemble themselves.

Organisms are described as "machines" to describe the idea that the properties of organisms are due to the interactions of their component parts, in just the same way that the properties of a motorcar result from the interactions between the parts that it contains. This approach to thinking about organisms is called "hierarchical reductionism" and the important aspect of this definition is its hierarchical nature. The properties of a complex system are explained in terms of the interactions between the parts which form the next level of complexity down in the

---

BASIC PRINCIPLES OF MOLECULAR BIOLOGY

THE MODERN VIEW OF LIFE
Organisms are self-assembling chemical machines programmed by their genes

1. All the properties of living organisms result from the properties of the proteins they contain

2. Each protein consists of one or more linear chains of amino acids and its properties depend on the sequence of these amino acids

3. The amino acid sequence in each chain of a protein is determined by the sequence of bases in the gene encoding that chain

4. The genes in each organism are handed on to the next generation

---

Fig. 4.3

hierarchy of complexity, and this process is continued down through decreasing levels of complexity. In other words, it is much more useful to describe how a motor car works in terms of the interactions between the cylinders, petrol tank, carburettor etc., than in terms of fundamental particles like electrons and quarks, even though it is the properties of fundamental particles that ultimately account for everything.

If we apply this reductionist approach to organisms, we discover that organisms are composed of several different levels of complexity, from organs to tissue, from tissues to cells, and from cells to subcellular organelles. But the ultimate components – the parts that determine the interactions at all the higher levels of organisation – are protein molecules. This simple fact enables us to construct the set of four principles shown in Fig. 4.3 that form the essence of molecular biology.

Principle 1: All the properties of living organisms result from the properties of the proteins they contain.

Proteins are the action molecules of all organisms, that is, they carry out most of the processes necessary to sustain the living state. Proteins function by presenting binding sites on their surface for chemicals in their environment, including other proteins – proteins are chemicals too! These sites are able to recognise and bind a huge range of chemicals in a highly specific fashion. Specificity means that each type of protein binds just one type of chemical, so cells contain many different proteins because cells are composed of many different chemicals. Among the many functions of proteins is to act as enzymes. Enzymes are defined as specific catalysts, that is, each type of enzyme catalyses one type of chemical reaction, which would proceed very slowly, or not at all, in its absence. The word "enzyme" originates from the Greek words for "in yeast" because the early work on enzymes was carried out by German scientists in the 1890s with extracts of yeast. Each enzyme has its own name, which indicates the type of reaction that it catalyses. For example, hexokinase catalyses the transfer of a phosphate group from a donor molecule called ATP to a carbon atom at position six in glucose to give glucose-6-phosphate.

Because each organism carries out several thousand different kinds of chemical reaction, the total constituting what is called metabolism, it follows that each organism contain thousands of different kinds of enzyme. When we add to their roles as catalysts the many other functions of proteins, such as signalling molecules, structural components and defence agents, it is clear that proteins are the most sophisticated group of chemicals known. So the next question is obvious – what determines the specific properties of all these different proteins?

Principle 2: Each protein consists of one or more linear chains of amino acids and its properties depend on the sequence of these amino acids.

Proteins are linear polymers of twenty different kinds of amino acid strung together like beads on a chain. Each chain is called a polypeptide and has a unique sequence of amino acids. Because there are twenty different kinds of amino acid found in proteins, and the length of chains can be anywhere between about 50 and 500 amino acids long, the total possible number of unique chains is very large. If we take the average length of a polypeptide chain as 300 amino acids, there are 20^{300} possible sequences. This number is much larger than the estimated total number

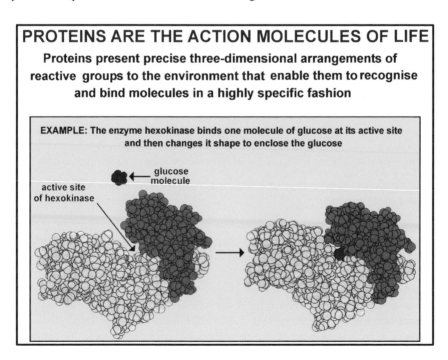

Fig. 4.4

of fundamental particles in the entire observable universe ($\sim 10^{80}$). So many more different kinds of protein are possible in principle than have actually appeared in evolution. We now have the genetic engineering technology to make any of these novel proteins, and some of them may well turn out to be useful.

How does the unique sequence of amino acids in the chains of a given protein determine the specific properties of that protein? The answer is that this sequence determines how each chain folds into a specific three-dimensional shape, called its conformation. Figure 4.4 shows a model of the specific conformation of one molecule of the enzyme hexokinase.

Each sphere in this model represents the size and position of an atom of either carbon, oxygen or nitrogen; hydrogen atoms, of which there are many, are omitted for clarity. The conformation of hexokinase is unique to that enzyme, and each molecule of hexokinase is identical. The structure contains an active site which binds a glucose molecule by interactions between the chemical groups of the glucose molecule and the chemical groups of the amino acids that line the active site of the enzyme. The way in which a chemical fits into an active site is sometimes described as analogous to the way a key fits into a lock, in the sense that there is a complementarity of surface features.

The conformation unique to hexokinase is formed by specific interactions between the side chains of the different amino acids along the polypeptide chain. What interactions are possible, and hence the shape of the folded molecule, is determined solely by the sequence of amino acids. Once that sequence is specified and synthesized, the chain folds spontaneously into its functional conformation. This fact is sometimes called the "principle of self-assembly" because all the information for the conformation is contained within the sequence. Modern techniques enable us to determine the sequence of amino acids in any polypeptide chain, but, as yet, it is not possible to predict how a given chain will fold. A major goal of protein chemists is to determine the rules of folding so that we can make novel proteins with useful properties.

Principle 3: The amino acid sequence in each chain of a protein is determined by the sequence of bases in the gene encoding that protein.

The amino acid sequences of proteins are determined by other sequences written in a different chemical language – the sequences of bases in nucleic acids. Nucleic acids are linear polymers of four different kinds of base and two different kinds of sugar phosphate. The two types of sugar phosphate define the two types of nucleic acids; ribose defines ribonucleic acid or RNA, while deoxyribose defines deoxyribonucleic acid or DNA. Although both proteins and nucleic acids are linear polymers, they differ greatly in the sizes of the individual molecules. In the case of proteins, the functional units are the individual molecules, such as hexokinase, but in the case of DNA, the functional units, or *genes*, are joined together to form giant molecules consisting of millions of bases and sugar phosphates. RNA is more like protein however, consisting of hundreds to thousands of bases and sugar phosphates in each molecule. This difference in size between molecules of DNA and molecules of RNA reflects their different functions – DNA encodes the genetic information that is handed on during cell division while RNA is part of the mechanism by which that genetic information is expressed. The only exceptions are some viruses that use RNA, not DNA, as their genetic material.

How a gene determines the amino acid sequence of a protein chain is the most complicated biochemical process yet discovered. This complexity is required because one chemical language – the base sequence of DNA – has to be translated into another chemical language – the amino acid sequence of protein chains. It is important to realise that this process does not involve a conversion of the bases themselves into amino acids. What flows from bases to amino acids is sequence information, not material.

The mechanism of gene expression is outlined in Fig. 4.5. DNA acts as a template that allows the sequence of bases in one strand of the double helix of each gene to be copied in the form of RNA. This process is called *transcription*, and produces from each gene many copies of a RNA molecule called messenger RNA (mRNA). Each molecule of mRNA is the about same length as the gene that acts as a template for its synthesis, and the enzyme that carries out this process is called RNA polymerase. The order of bases in mRNA is the same as the order of bases in one strand of the double helix of DNA from which it is copied. The mRNA in turn acts as a template for a second process called *translation,* in which amino acids are joined together in a linear order determined by the order of bases in the mRNA. Each molecule of mRNA can programme the synthesis of many molecules of the protein chain that it encodes. The rules governing which bases determine which amino acids are joined together constitute the *genetic code*.

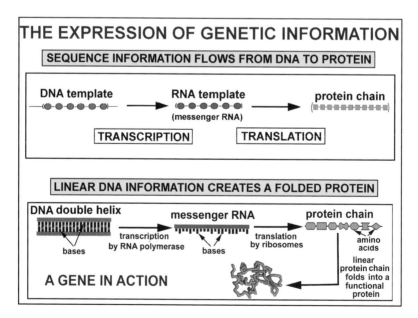

Fig. 4.5

You may like to think of gene expression in terms of computer terminology. The nucleic acids represent the software – the memory that contains the information to make proteins. The proteins are the hardware –they constitute the physical apparatus that executes the programme that stored in the software. You may have heard media people say that DNA is the "blueprint" for each organism, but this is incorrect. The term "blueprint" is traditionally used for a physical description of the parts of a given structure, such as a ship or a camera, but DNA contains no such description of an organism. A better analogy is that DNA is a recipe – it contains the information to make all the proteins of the organism, and it is the interaction of these proteins that produce the structure of the organism, just as it is the mixture of flour, butter, raisins etc that makes the structure of a cake.

The process of translation requires many interacting components. Ribosomes, you will remember, are the structures inside each cell that decode the base sequence information in each molecule of messenger RNA. By "decode", we mean that the ribosome is essentially a device for converting the base sequence information in each messenger RNA into the amino acid sequence information of the protein encoded by each gene, just as during the Second World war, the scientists at Bletchley Park in England built a machine that was able to decode the secret messages sent out by German military forces. Recall that each gene is defined by its unique sequence of bases, while each polypeptide is defined by its unique sequence of amino acids. Thus ribosomes use the base sequence in messenger RNA to join amino acids together in the correct order.

Ribosomes are complex organelles consisting of 50–70 protein molecules bound to 2 or 3 RNA molecules, depending on the species. A human liver cell contains several million ribosomes. Each ribosome consists of two subunits, called large and small. These subunits are separate when they are not synthesizing protein chains, but are joined together when the ribosome attaches to one end of a mRNA molecule. The ribosome then moves along the mRNA molecule, joining amino acids together as it goes, in the order specified by the order of bases in the mRNA. Several ribosomes can attach to a single molecule of mRNA, forming a structure called a polyribosome, or polysome. The existence of polysomes allows many copies of the same protein chain to be made rapidly. Figure 4.6 shows the front and back ends of a polysome isolated from the salivary glands of an insect; the middle section on the left hand side is missing.

Why is decoding by ribosomes absolutely vital for life? It is vital because it is the sequence of amino acids in each linear chain of protein that determines how that chain folds into the compact three-dimensional structure that enables that protein to carry out its specific function. Remember that it is proteins that are the "action molecules" of the organism – that is, the molecules that carry out all the thousands of different functions necessary for life. Genes by comparison, are inert and rather boring – all they do is contain the sequence codes for proteins; some genes encode RNA molecules that are not messenger RNAs to make proteins, but RNAs that either regulate transcription and translation or form part of the structure of ribosomes themselves. You might like to think of this arrangement by analogy with a tape recorder. The tape is like the DNA; it is essential because it contains the information

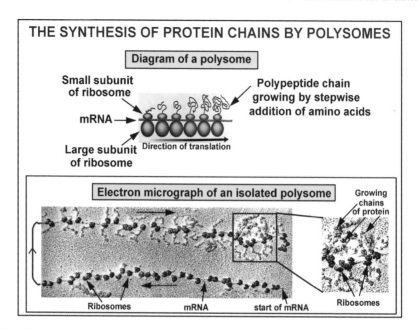

Fig. 4.6

to make the music, but in itself, it is uniform, rather dull and doesn't do anything on its own. The music produced by decoding the linear information on the tape, on the other hand, is much more complex in structure and much more interesting – in organisms, the proteins are the music.

The above simple description of gene expression omits an important aspect of the genome – that gene expression is regulated in precise ways, both in terms of how active a given gene is at a particular time, and where it is active in the body of muticellular organisms. By active, we mean that the gene is being transcribed into RNA. So genes can be switched on and off, and the switches are proteins that bind to particular regions of the gene being regulated. We call these proteins *transcription factors* for obvious reasons.

It might seem to be a paradox that organisms show high diversity at the anatomical level but high similarity at the molecular level – the so-called "Unity of Biochemistry" referred to earlier in this section. The explanation of this apparent paradox is that much of the diversity at the anatomical level is due to how some genes are regulated, and not in the genes themselves. By gene regulation remember, we mean where in the body, and when in the life of the organism, particular genes are transcribed. It is now clear that much of evolution is at the level of gene regulation rather than at the level of the function of the gene product.

Let me give you an example. About 15,000 years ago, the ancestors of stickleback fish lived exclusively in the oceans. But at the end of the last Ice Age many stickleback populations found themselves in newly formed freshwater lakes. Today, sticklebacks that live in the ocean have pelvic spines that offer some protection against animals that prey on them by making them too big and prickly to swallow. But sticklebacks that live in shallow freshwater lakes today lack such spines because the larger predator problem is much less in that environment, and the spines are a disadvantage against another predator found only in freshwater - dragonfly larvae that grasp the spines. Sticklebacks from the ocean will reproduce with sticklebacks from freshwater lakes so experiments are possible to determine the genetic difference between the two types. Molecular techniques have been used to identify a gene that affects where in the body of the fish the genes for these spines are expressed. This regulatory gene is normally expressed in the head, trunk, pelvis and tail because the products of the genes that it regulates are required for other processes in these regions, but in the freshwater fish, the gene is no longer expressed in the pelvis. So the change in anatomy has been produced by changes in gene regulation and not in the genes determining the structure of the spines.

Principle 4: The genes in each organism are handed on to the next generation.

   In animals such as humans, each generation is formed from a zygote, a cell formed by the genetic material of a sperm entering and fusing with the genetic material in an ovum. All the information to make another human is contained within this zygote, but it is a common misconception to conclude that all that is inherited is DNA. The DNA contains all the information to make all the proteins required to construct an organism but this information requires a pre-existing cell in which it can be utilised. Life is a cellular phenomenon, and so the continuity of life from its origin some four billion years ago resides in cells, not in DNA. This conclusion is summarized in the adage "All cells from cells".

The Tree of Life

I said earlier that all organisms are genetically related. An example of a gene sequence that is highly conserved in all forms of life is shown in Fig. 4.7. This diagram compares the amino acid sequence of part of a protein involved in controlling protein synthesis by ribosomes in organisms from all three domains of life. This protein is called an elongation factor, because without it the addition of amino acids by the ribosomes to the growing chains of protein stops. Each amino acid (there are twenty different ones in proteins) is represented by a letter e.g. A stands for alanine. You can see that there are several regions where the sequence of amino acids is identical between very different species – from humans to bacteria.

   There is fossil evidence that bacteria have been on the Earth for at least 3 billion years, so the interpretation is that these sequences have been conserved over that huge length of time because they are essential for the elongation factor to help the ribosome to carry out its job of making proteins. Because the ribosome is such a vital component of all cells, its structure is highly conserved in all organisms. This is true for both the protein and the RNA molecules that make up the ribosome. So it is not surprising that the base sequence of the RNA component of the ribosome is also highly conserved in all organisms – too much variation would run the risk that

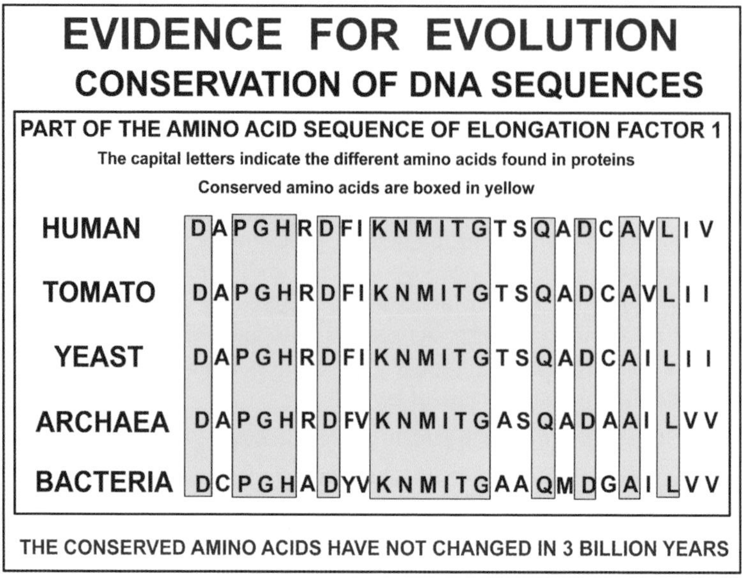

EVIDENCE FOR EVOLUTION
CONSERVATION OF DNA SEQUENCES

PART OF THE AMINO ACID SEQUENCE OF ELONGATION FACTOR 1

The capital letters indicate the different amino acids found in proteins

Conserved amino acids are boxed in yellow

HUMAN   D A P G H R D F I K N M I T G T S Q A D C A V L I V

TOMATO   D A P G H R D F I K N M I T G T S Q A D C A V L I I

YEAST   D A P G H R D F I K N M I T G T S Q A D C A I L I I

ARCHAEA D A P G H R D F V K N M I T G A S Q A D A A I L V V

BACTERIA D C P G H A D Y V K N M I T G A A Q M D G A I L V V

THE CONSERVED AMINO ACIDS HAVE NOT CHANGED IN 3 BILLION YEARS

Fig. 4.7

the ribosomes will not work well enough at making proteins. This constraint means that the sequence of ribosomal RNA changes only very slowly during evolution, and so can be used to determine how the major groups of organisms are related to one another.

Figure 4.8 shows the three-domain classification scheme that is currently a popular model for describing the relations between the major groups of living organism on the Earth. This model is derived by comparing the base sequence of the genes encoding the RNA component of the small subunit of ribosomes in the different organisms that exist today – these are called extant organisms. The more similar these sequences, the closer together the organisms are placed. This scheme is often described as a family or phylogenetic tree, that is, a diagram that shows how organisms are genetically related to one another. The metaphor of a tree to describe the relatedness of organisms was not invented by Charles Darwin but was promoted by him. In Chapter 4 of *On the Origin of Species* he states:

> The affinities of all beings of the same class have sometimes been represented by a great tree.... The green and budding twigs may represent existing species: and those produced in each former year may represent the long succession of extinct species. At each period of growth all the growing twigs have tried to branch out on all sides, and to overtop and kill the surrounding twigs and branches, in the same manner as species and groups of species have tried to overmaster other species in the great battle of life.

You will note that all the branches in the tree shown in Fig. 4.8 are connected to one another and all converge on the same point at the bottom. So this tree is saying

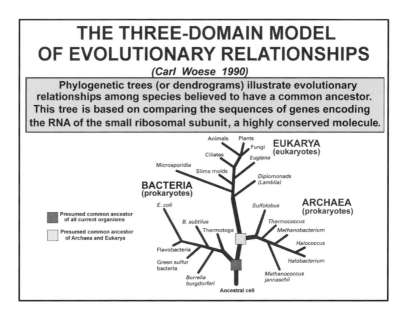

Fig. 4.8

that all organisms are derived from one common ancestor and are related to one another – the basic predictions of Darwin's evolutionary theory.

Figure 4.8 contains some similar terms that are often confused, even by some biologists. The essential point to remember is that the terms "prokaryote" and "eukaryote" describe basic structural differences between two types of cell, while the terms "Archaea", "Bacteria" and "Eukarya" describe evolutionary relationships as deduced from DNA sequences. Eukaryotic cells in general are larger and more complex than prokaryotic cells, and contain a number of internal organelles, such as a nucleus, mitochondria, lysosomes, endoplasmic reticulum, Golgi bodies and a complex cytoskeleton, which are not found in prokaryotic cells. Plant cells contain, in addition, structures termed plastids, of which the chloroplast is the most studied example. Figure 4.9 illustrates these striking structural differences between prokaryotic cells and eukaryotic cells.

Eukaryotic cells have their DNA separated from the cytoplasm by a nuclear membrane, but prokaryotic cells have their DNA in direct contact with the cytoplasm. The cytoplasm is defined as everything outside the DNA. This difference affects the relation between transcription and translation, because transcription takes place only in the nucleus of eukaryotes, while translation takes place only in the cytoplasm. In contrast, transcription and translation take place in the same compartment in prokaryotic cells, and so these processes can be coupled together – a messenger RNA molecule can be translated into protein while it is still being synthesized from a DNA template. Most prokaryotes are unicellular, and so the close coupling of transcription with translation in the same compartment allows these cells to grow

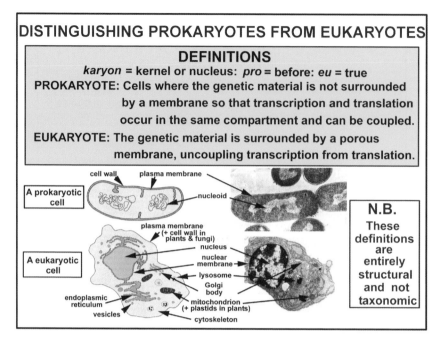

Fig. 4.9

quickly when they encounter favourable growth conditions, and so outcompete other organisms. In contrast, most eukaryotes do not compete at the level of growth and cell division, but by making more elaborate internal compartments that allow the cells to become larger and more structurally sophisticated than prokaryotic cells. This internal specialisation allows eukaryotic cells to form multicellular organisms where different cells have different functions. All the organisms visible to the naked eye are eukaryotes.

Both the Archaea and the Bacteria are made of prokaryotic cells, while all the Eukarya are made of eukaryotic cells. Confusion has arisen between these terms because, while the terms "prokaryote" and " eukaryote" were first used to describe the main structural difference between the two types of cell, they were then assumed by some biologists to also reflect their evolutionary relationships. But according to the three-domain model presented in Fig. 4.8, the prokaryotic Archaea are more closely related to the eukaryotic Eukarya than they are to the prokaryotic Bacteria. This tree is based on the analysis of one gene – the gene that encodes the rRNA found in the small ribosomal subunit, so we need to ask the question as to whether trees based on similarities in other genes look the same. You will also note that this tree is based on the assumption that different species are related by vertical descent, that is, genes are transferred from parent cells to daughter cells only by the processes of cell division and sexual reproduction. But suppose that there are other ways than vertical descent by which genes transferred between cells, what happens to the tree diagram then? The three-domain tree also supposes that eukaryotes arose from prokaryotes in some way, so we need to ask how this might have happened. Did some prokaryotes develop eukaryotic features on their own or did some prokaryotic cells fuse together to produce the larger, more complex type of cell?

Lateral Gene Transfer

Evidence has accumulated since the 1970s for a widespread process by which genes are transferred between unrelated organisms. This process is called lateral gene transfer (abbreviated to LGT), also known as horizontal gene transfer (abbreviated to HGT).

Lateral gene transfer (LGT) is defined as any process in which an organism incorporates genetic material from another organism without being the direct descendent of that organism. All definitions of the term "species" assume that an organism gets all its genes from one or two parents which are very like that organism, but the occurrence of lateral gene transfer makes this assumption false in some cases, especially in prokaryotes.

LGT was first described by Japanese scientists in 1959, when it was found that resistance to some antibiotics was transferred between different species of bacteria, but the significance of this phenomenon was not appreciated by Western scientists until the 1970s. You will be familiar from media reports with the current problem of the transfer of antibiotic resistance between harmless bacteria and those that that cause life-threatening conditions in humans, especially in hospital environments. We now know that LGT is common amongst bacteria, including ones that are only distantly related, and several different mechanisms have been discovered. The simplest mechanism is called *transformation,* where bacteria and archaea take up foreign DNA from the environment. Most of this DNA is degraded within the cell, but some becomes incorporated into the host chromosome and may confer new properties, such as resistance to antibiotics, on the cell that enhance survival in certain environments. A different mechanism is called *transduction,* and involves transfer of DNA by viruses that infect bacterial cells but do not destroy them. A third process involves direct contact between a cell that donates some DNA and a different cell that receives this DNA, and is called *conjugation.* LTG has also been observed between some bacteria and some archaea, between some bacteria and some fungi, and between some bacteria and some unicellular eukaryotes.

In multicellular eukaryotes, the major cause of LGT is the phenomenon of *endosymbiosis.* Endosymbiosis is now defined generally as one type of cell living inside a cell from an unrelated species without harming it, but was initially defined as the evolutionary process by which the plastids and mitochondria of eukaryotic cells originated from free-living bacteria that were taken up by other cells. The engulfed bacteria then evolved into symbiotic relationships within the host cell to form plastids and mitochondria that are now totally dependent on their host cell for their continued existence, because most of the genes for the proteins in these organelles have been laterally transferred to the nucleus. "Plastid" is a general term used to describe a type of organelle found only in plants; the best-studied type of plastid is the chloroplast, the green organelle that carries out photosynthesis.

Mitochondria originated from a group of bacteria called the α-proteobacteria, while plastids originated from cyanobacteria, so-called because of their photosynthetic pigments. Thus the appearance of the vital processes of aerobic respiration and photosynthesis in eukaryotes is due to endosymbiosis and massive lateral gene transfer from the ingested bacteria to the nucleus of the host cell. But the phenomenon of endosymbiosis also includes many other cases of bacteria that have taken up residence inside eukaryotic cells, where they both enjoy and confer some metabolic benefit, but have not evolved into either mitochondria or plastids. For instance many insects that live on plants, such as aphids, contain bacteria in their cells. These bacteria provide the insects with certain compounds that are lacking in the sap that these insects extract from the plants. There are even cases of endosymbioses where eukaryotic cells have taken up residence inside other eukaryotic cells.

The evidence for endosymbiosis is summarized in Fig. 4.10. A German botanist called Schimper suggested in a footnote of a paper published in 1883 that maybe chloroplasts originated from cyanobacteria because they looked similar, and moreover he could see chloroplasts dividing inside plant cells, as though they were cells themselves. This idea was developed much further in 1905 by the Russian biologist Mereschkowsky, who stressed not only that chloroplasts were seen to divide, but that they continuing dividing and carrying out photosynthesis in parts of plant cells from which the nucleus had been removed by dissection. This latter observation implied that chloroplasts were independent of the nucleus to some extent, as might be expected if they originated from free-living cyanobacteria. Today we know that chloroplasts have only a very limited independence from the nucleus because most of the genes for the three thousand or so proteins that make up chloroplasts are encoded in the nucleus. The chloroplast does contain about one hundred genes, the exact number depending on species, and it does make some of its own proteins using its own ribosomes, but the vast majority of chloroplast proteins are made by cytosolic ribosomes and then transported across the chloroplast envelope. So chloroplasts do not divide and photosynthesize when isolated from plant cells for more than a few hours. It follows that during evolution there has been a massive lateral transfer of genes for chloroplasts and mitochondrial proteins from the original endosymbionts to the nucleus. An American scientist called Ivan Wallin extended the endosymbiont idea to the origin of mitochondria in 1923.

Like many scientific ideas that later turn out to be correct, this suggestion by Schimper, Mereschkowsky and Wallin was not accepted for a long time. Most biologists found it hard to swallow the idea the chloroplasts and mitochondria could have descended from free-living cells that had somehow ended up inside larger cells. It was research in the 1960s showing that both mitochondria and chloroplasts contain their own complete genetic systems, separate from that in the nucleus and cytosol, that revived the idea. A genetic system is defined as one that contains both DNA with protein-encoding genes, the enzymes to transcribe these genes into messenger RNAs, and the ribosomal translation apparatus that uses these messenger RNAs to

THE ENDOSYMBIONT HYPOTHESIS
Proposed by Mereschkowsky, 1905: revived by Margulis, 1967

PLASTIDS AND MITOCHONDRIA RESEMBLE BACTERIA

1. Both arise from pre-existing organelles by binary fission.
2. Both contain circular DNA encoding proteins & ribosomal RNA.
3. Both translate using ribosomes sensitive to chloramphenicol.
4. Both are surrounded by a double membrane.

EUKARYOTIC GENE SEQUENCES FALL INTO THREE GROUPS

1. Plastid genes are most closely related to cyanobacterial genes.
2. Mitochondrial genes are most closely related to bacterial genes.
3. Nuclear genes fall into three groups - bacterial, archaeal & unique.

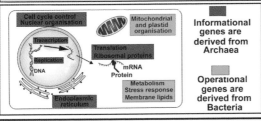

Informational genes are derived from Archaea

Operational genes are derived from Bacteria

CONCLUSION

Eukaryotic cells are a cut-and-paste job
(Wikipedia reference: `Endosymbiotic theory`)

Fig. 4.10

synthesize proteins. Both mitochondria and chloroplasts contain their own DNA and their own ribosomes, which in their detailed molecular properties resemble those from proteobacteria and cyanobacteria, but differ from the DNA and ribosomes in the nucleus and cytosol. Like chloroplasts, mitochondria contain the genes for less than 1% of all the organellar proteins, the vast majority residing in the nucleus. Thus plant cells are more complex than animal cells, in the sense that they all contain three genetic systems, whereas animal cells contain only two. It has also been established that mitochondria and chloroplasts are inherited directly from the parent cell in the form of smaller membrane-bound structures, termed promitochondria and proplastids. This fact recalls the adage "All cells from cells", referred to in the section entitled *The Principles of Molecular Biology*.

The Origin of Eukaryotes

You will note from Fig. 4.10 that comparison of whole genome sequences from Bacteria, Archaea and Eukarya show that eukaryotic genomes contain some genes more closely related to those found in Bacteria today, and other genes related to those from Archaea today. Most interestingly, these genes can be largely grouped into two distinct classes, depending on the functions of the proteins they encode. Thus the archaeal-related genes encode proteins involve in informational processes, that is, processes such as DNA replication, transcription and translation. The bacterial-related genes, in contrast, encode proteins involved in metabolic processes such as respiration and photosynthesis – operational processes. The startling conclusion is that eukaryotes are descended from hybrids of Archaea and Bacteria – all life is microbial!

There is much current debate as to precisely how archaeal cells combined with bacterial cells to produce eukaryotic cells, but no general consensus has been reached as yet. One recent model proposes that eukaryotic cells originated when an archaeal cell engulfed bacterial cells that evolved into mitochondria, while plastids originated by a later engulfment of cyanobacteria by these eukaryotic cells (Fig. 4.11). There is no experimental evidence as yet to show that such a process of engulfment can occur today.

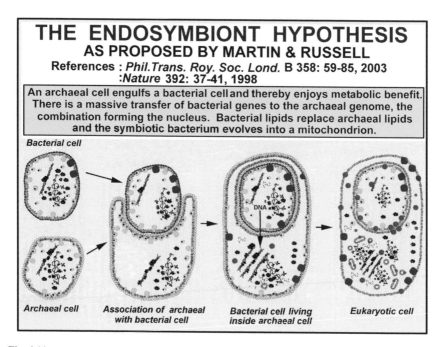

Fig. 4.11

The fossil record suggests that prokaryotic cells existed at least 3500 million years ago, but that eukaryotic fossils did not appear until around 2000 million years later. This long gap has prompted some biologists to suggest that the origin of eukaryotic cells is a highly improbable event, and thus if life does exist on planets outside the Solar System, it is likely to be only prokaryotic in nature. It would be disappointing if the Earth is the only abode of eukaryotic life in the entire Universe, since intelligent animals like ourselves are eukaryotes. On the other hand, if this view is correct, at least we do not have to worry about invasion by advanced aliens!

The discovery of lateral gene transfer is changing the way that we think that organisms are related to one another. We can no longer assume that all the genetic information found in each species has originated from the direct ancestors of that species. It follows that the tree of life metaphor favoured by Charles Darwin is being modified by the overlaying of the branches of the tree by interconnections that represent lateral gene transfer events, including those that accompany endosymbiosis. Figure 4.12 is an attempt to illustrate this new view of evolutionary relationships.

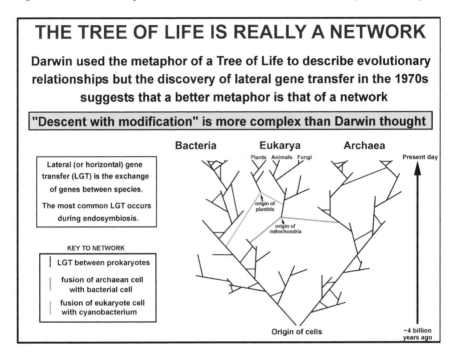

Fig. 4.12

Similarities at the Anatomical Level

There are many examples of similarities between different organisms at the anatomical level. Figure 4.13 compares the forelimbs of seven different tetrapod (four-legged) animals. These animals use these limbs for different purposes – hopping, running, flying, walking and swimming. There is no obvious functional or environmental reason why these forelimbs should all have five digits rather than three or seven, but they do, and more amazingly, they also share a common pattern of anatomy.

This fact impressed Darwin. In his book "*On the Origin of Species*", he says:

> What could be more curious than that the hand of man, formed for grasping, that of a mole for digging, the leg of a horse, the paddle of a porpoise and the wing of a bat, should all be constructed in the same pattern and should include similar bones in the same relative positions? Why should similar bones have been created in the formation of a wing and a leg of a bat, used as they are for totally different purposes?

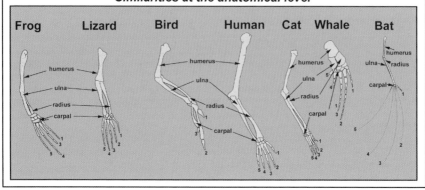

Fig. 4.13

The argument here is that a human designer who was trying to make the best possible wing and the best possible leg would have no need to ensure that the basic components of these two structures were the same. If he or she did, this would be a serious limitation on their attempt to produce the best design. Evolution by natural selection on the other hand, has no knowledge of the best design – it simply selects from whatever is available at the time for anything that enhances survival and reproduction in the current environment. Thus the basic structures inside tetrapod forelimbs are similar because they are related by descent, as predicted by evolutionary theory, not because they represent the best design.

Anatomical similarities can also be observed during the development of some animals which differ greatly in their adult forms. Such differences reflect evolutionary changes in the later developmental processes that produce the adult forms, while the developmental processes that produce the immature forms are more similar because they are evolutionarily related. Figure 4.14 shows that the larval stages of three types of invertebrate are more similar to one another than are the adult forms.

Religious people who dislike the idea of natural selection sometimes argue that the similarities between forelimbs illustrated in Fig. 4.13 might reflect the whim of a supernatural creator. The problem with this type of argument is that it can be used to explain whatever type of structures are found inside these forelimbs, so it has no predictive power. The creator might even prefer to create forelimbs that are not

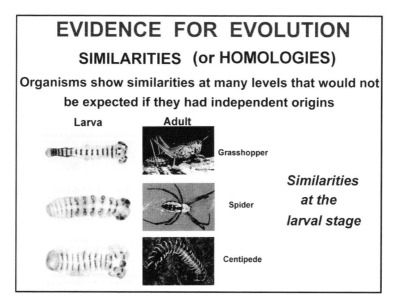

Fig. 4.14

the best design possible. Because there is no way to determine the preferences of a hypothetical creator, the supernatural explanation lacks the predictive power of the natural explanation. This is an example of my earlier suggestion that the essential difference between science and religion is that the latter lacks a methodology by which its claims can to be tested.

Direct Observation

A strong line of evidence for evolution is that it can be observed for some species, both in the field and experimentally (Fig. 4.15).

We distinguish two levels of evolution. *Microevolution* is defined as evolution below the species level, and *macroevolution* as above the species level, but this distinction is only for convenience – both processes are part of a continuous spectrum of change across differing time scales, and different techniques are used to study them. For some organisms that reproduce rapidly, such as viruses, bacteria and some birds and insects, changes in their genetic composition from generation to generation can be observed directly. This is the case for example with the HIV virus that causes AIDS in humans – the evolution of this virus in an infected person treated with antiviral compounds can be seen in just three weeks.

There are two reasons for the rapid evolution of the HIV virus. Firstly, the invasion of a single white blood cell by a single virus particle results in the production of about 10,000 new virus particles every 24 h by that cell. Secondly, the mutation rate (the rate of change of the base sequence of the genetic material) is high enough that on average each new particle carries one mutation. A small minority of these mutations cause the virus to become resistant to the antiviral drugs being used. Thus the resistant variants increase with time in the presence of the drugs. The result is that the drugs rapidly become less effective with time, making this disease difficult to treat.

Evidence for evolution is also seen from the domestication of plants and animals by humans in the last ten thousand years. In the first chapter of his famous book, Charles Darwin talked about the evolution of domestic varieties of animals such as dogs and pigeons from wild ancestors. Darwin was especially interested in the way that selective breeding by pigeon fanciers has produced all the varieties of pigeon that they show at exhibitions. All the breeds favoured by pigeon fanciers have been bred from the wild rock pigeon – Fig. 4.16 shows a few examples. Dogs have been evolving from wolves for at least ten thousand years, and possibly for ten times longer. Selection of favourable traits by humans has produced the 300 or so varieties of dog that we see today. Figure 4.16 shows just two modern varieties of dog, compared with their remote ancestor.

The domestic plants that we all rely upon for food have also been produced by breeders selecting those variants that possess desired characteristics – desired that is by humans, not by adapting to the natural environment. Figure 4.16 shows the different crop plants that have been bred from the wild cabbage by plant breeders. All these are examples of microevolution. This process produces new varieties, but these varieties are still the same species. So for instance modern dogs can, and do, interbreed with wolves.

EVIDENCE FOR EVOLUTION

MICROEVOLUTION CAN BE OBSERVED DIRECTLY

Change above the species level is called macroevolution and is much too slow to be observed directly. But change within a species, or microevolution, can be observed for rapidly reproducing organisms. Rates of microevolution in both the laboratory and the field are more than enough to account for the rates of change seen in the fossil record

EXPERIMENTS CAN PRODUCE MICROEVOLUTION

Almost all our domestic and agricultural plants and animals have been produced by artificial selection

Fig. 4.15

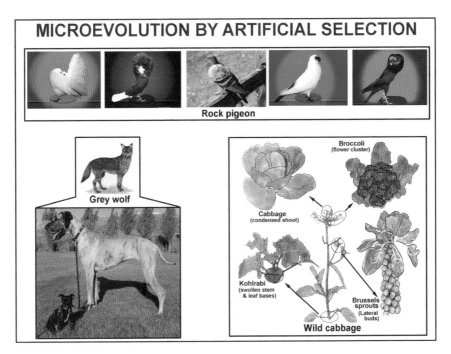

Fig. 4.16

Transitional Fossils

Unlike microevolution, macroevolution is much too slow to be seen directly but its occurrence is inferred – just as electrons cannot be observed directly but their existence is inferred. Macroevolution is inferred from the fossil record and from the principle of uniformitarianism (Fig. 4.17). This principle was formulated by the Scottish geologist James Hutton in the late eighteenth century. He proposed that rock is being continually eroded and washed down into the seas, where it sediments into layers and is compressed back into rock. This sedimentary rock is then uplifted out of the water by earthquakes so that the erosion cycle is repeated. The observed thickness of sedimentary rocks suggested that this process has been continuing for very long periods of time i.e. it was a "uniform" process. This idea led to the gener-alisation that the continued operation of processes observable today could account for the geology of the planet if prolonged over millions of years.

Figure 4.17 defines this principle – the idea that the present is the key to the past. This view was very influential in the development of both biology and geology in the eighteenth and nineteenth centuries. This principle stems from the naturalistic view that the world is governed by unvarying regularities, and is also an exam-ple of Occam's razor in action, because it is based on the simplest hypothesis – that processes acting now also operated in the past. But the principle does not rule

EVIDENCE FOR EVOLUTION

THE PRINCIPLE OF UNIFORMITARIANISM

Definition

Uniformitarianism is the assumption that natural processes
that are observed to be operating in the present
also operated in the past.

This principle was popularised by the British geologists James Hutton
and Charles Lyell in the late 18th and early 19th centuries, but is now
used in other sciences whose objects of study are in the past, such as
astronomy, palaeontology and biology. Charles Darwin was influenced
by the geology book written by Charles Lyell while on his voyage
around the world in HMS *Beagle* during 1831 - 1836.

"THE PRESENT IS THE KEY TO THE PAST"

Fig. 4.17

out sudden catastrophic events occurring against this uniform background of geological change. There is evidence of at least two major extinction events linked to extreme volcanic activity and asteroid strikes that caused the Earth's atmosphere to darken for periods long enough to render extinct many organisms that rely, directly or indirectly, on photosynthesis.

Macroevolution is also inferred from the fossil record. If some species can change into other species over long periods of time, there might to be fossils of transitional forms between them – and there are! Transitional species are defined as those that show a mixture of features from both their ancestors and their descendents. We now have transitional fossils between fish and amphibians, reptiles and birds, hippos and whales, and apes and humans. Figure 4.18 shows some examples of fossils sharing characteristics of both dinosaurs and birds.

On the left of Fig. 4.18 is shown one of the famous *Archaeopteryx* fossils, of which ten specimens are now known, all found in southern Germany. The first was found in 1861, just two years after the publication of *On the Origin of Species*, and the last in 2005. These fossils were all preserved in limestone in Bavaria, and date to the Jurassic period, about 150 million years ago. The word *Archaeopteryx* means "ancient wing".

These fossils show a mixture of avian and dinosaur traits. They share with dinosaurs, jaws with sharp teeth, three fingers with claws, a long bony tail and an extensible second toe. They share with birds a wishbone, flight feathers and wings.

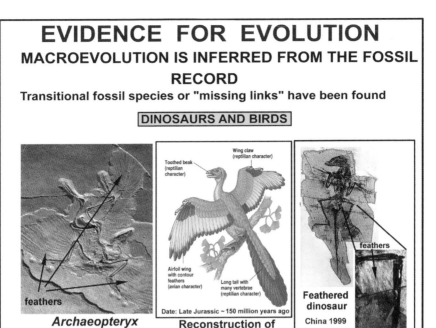

EVIDENCE FOR EVOLUTION

MACROEVOLUTION IS INFERRED FROM THE FOSSIL RECORD

Transitional fossil species or "missing links" have been found

DINOSAURS AND BIRDS

Wing claw (reptilian character)

Toothed beak (reptilian character)

Airfoil wing with contour feathers (avian character)

Long tail with many vertebrae (reptilian character)

feathers

feathers

Date: Late Jurassic ~150 million years ago

Archaeopteryx
First found in Germany in 1861.
Nine similar fossils were found later.

Reconstruction of
Archaeopteryx

Feathered dinosaur
China 1999

Fig. 4.18

The flight feathers are well developed and asymmetrical, like those of modern birds. *Archaeopteryx* was probably a glider rather than a flapping animal because it lacks a large breastbone required for the anchoring of the powerful muscles necessary for flight, and the anatomy of its shoulder suggests it was unable to lift its wings above its back. Reconstructions of its skull by computer tomography (CT scanning) of the skulls show that the regions of the brain concerned with vision were well developed. The structure of the inner ear closely resembles that of modern birds rather than that of reptiles. These observations are interpreted to indicate that *Archaeopteryx* had good vision, hearing and balance, but whether it lived as a tree-dwelling glider or evolved flight by running along the ground is the subject of continuing debate.

In China in the last 15 years, eight different feathered dinosaur fossils have been discovered (Fig. 4.18, right hand panel). From their anatomy, these were probably not capable of flapping flight, so their feathers may have evolved to act as insulation as these animals evolved warm-bloodedness, and only later were used for flight. This is a recurring theme in evolution – one thing leads to another. Natural selection is a very powerful opportunistic process – even a slight variation will be selected if it aids survival and reproduction, so traits selected for one purpose can form the basis for the subsequent development of quite different traits.

Figure 4.19 shows some fossils that bridge the gap between fish and tetrapods – animals with four limbs.

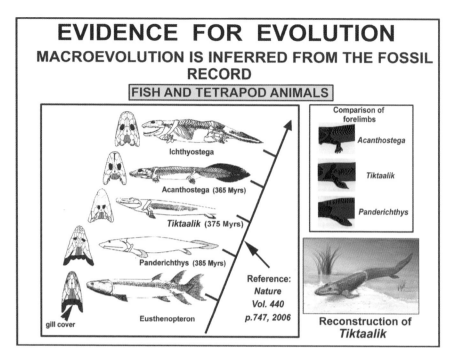

EVIDENCE FOR EVOLUTION
MACROEVOLUTION IS INFERRED FROM THE FOSSIL RECORD
FISH AND TETRAPOD ANIMALS

Ichthyostega

Acanthostega (365 Myrs)

Tiktaalik (375 Myrs)

Panderichthys (385 Myrs)

Eusthenopteron

gill cover

Reference: *Nature* Vol. 440 p.747, 2006

Comparison of forelimbs

Acanthostega

Tiktaalik

Panderichthys

Reconstruction of *Tiktaalik*

Fig. 4.19

The older fossils at the bottom show gill covers, fins and fish tails but no neck, while at the top, the gill covers have gone, the fins are now limbs with digits, there is a neck, but the fish tail is still present. In the middle is a fossil called *Tiktaalik,* an Inuit name reflecting that it was found recently in the Canadian Arctic. This find was reported in the journal *Nature* in 2006. *Tiktaalik* has fish scales, but no gill covers and the pectoral fins are almost, but not quite, tetrapod limbs – they still have fin rays. These fins would have allowed paddling, but they also have substantial bones that would have enabled *Tiktallik* to prop itself up in shallow water but not to walk.

A remarkable feature of the discovery of *Tiktaalik* is that it was not entirely accidental – the fossil hunters who found it were looking for it! By comparing the known fossils shown in Fig. 4.19 that link fish with tetrapods, they formed the hypothesis that the invasion of the land by vertebrate animals took place in a river environment some 375 million years ago. So they looked for fossils in rocks of this age whose geology suggested they had formed in a river delta. Looking for fossils requires unusual amounts of patience and it took five expeditions to Canada before they were successful. This remarkable story is related by Neil Shubin, one of these fossil hunters, in his book *Your Inner Fish* (see Further Reading). This is a good example of science in action – you identify a problem with existing knowledge, form a hypothesis about a possible solution and then seek observations or perform experiments to test that hypothesis.

When we call *Tiktaalik* a transitional species, this does not necessarily mean that this species was on the direct line of descent from fish to tetrapods – it could represent an extinct branch on the line of descent between these two groups. But *Tiktaalik* does demonstrate the past existence of a species intermediate in form between fish and tetrapods, as predicted by evolutionary theory. For this reason the term "transitional fossil" is better than the term "missing link". The value of such transitional fossils is that they show us the order of the evolutionary steps that connect one type of organism, such as fish, with a later type of organism, such as tetrapods. In the same way, *Archaeopteryx* probably is not on the direct line of descent of modern birds – remember that the tree of life is highly branched and most species of life are now extinct!

Logical Inference

Figure 4.20 illustrates another line of evidence in support of evolutionary theory – the fact that DNA undergoes mutation. Mutation is defined as a change in the base sequence of the genetic material of the organism. Mutation is the ultimate source of the variation between individuals in a population that was so well documented by Darwin in his book. Remember that Darwin knew nothing about the genetic material or the mechanism of heredity – he had some ideas about how the latter might work, but these were hopelessly wrong.

EVIDENCE FOR EVOLUTION
MUTATION IS THE SOURCE OF INHERITED VARIATION
Darwin's problem
Darwin knew neither the mechanism of heredity nor the the cause of variation. Today we know that inherited variation is caused by MUTATION - random copying errors during the replication of DNA.
(Compare with NATURAL SELECTION which is nonrandom)

TYPES OF MUTATION	RATES OF MUTATION		
		RNA virus	Bacteria to humans
1. Point mutation	Rate per base per replication	~1 in 10,000	~1 in 1,000,000,000
2. Slippage			
3. Transposition	Rate per genome per generation	1	Bacteria 0.001
4. Unequal crossing over			Humans 200
5. Chromosome deletion & duplication			

THE COMBINED EFFECTS OF MUTATION AND NATURAL SELECTION MAKE EVOLUTION INEVITABLE
THUS EVOLUTION IS BOTH A THEORY AND A FACT

Fig. 4.20

Unlike natural selection, which is a highly nonrandom process, mutation is a chance process – random copying errors that may occur every time DNA is replicated. The term "random" in this context means that the particular mutations that occur are unrelated to their effects on evolutionary fitness. A better term would be "accidental" or "undirected", but the term "random" is commonly used. This difference between the properties of natural selection and mutation is another common source of confusion in media debates about evolution.

So remember – *mutation is random, but natural selection is nonrandom.*

There are several different ways in which mutation can occur – these are listed on the left of Fig. 4.20. On the right, are some approximate rates of mutation.

RNA viruses have high rates of mutation because RNA, unlike DNA, is single-stranded and RNA viruses lack proofreading mechanisms. All cellular organisms contain proteins that are able to detect errors in base-pairing during the replication of DNA and correct them, using the information in the strand of DNA being copied. But RNA viruses lack such proofreading mechanisms, which is why the AIDS virus evolves so quickly – I mentioned earlier that you can observe the AIDS virus evolving within a single human individual in a few weeks. The appearance of the MRSA superbug in hospitals is another example of evolution in action – the selection here is created by our use of antibiotics.

Rates of mutation are expressed in several ways. The rate per base per replication is very low in everything except viruses. For example, every time a human cell divides about six new mutations arise on average, that is, six bases out of a total of six billion bases are changed. This seems a very small change, but when you consider the number of cell divisions required to make the gametes of an adult human, it turns out that the average person will accumulate in their gametes during their reproductive lifetime around 200 mutations. Most mutations turn out to be neutral in their effect, either because they occur in regions of the DNA that do not code for proteins or because they do not change the amino acid sequence. But a minority of mutations are harmful. For example, about 1 in 25,000 people are born with a single mutation in a gene that encodes a protein involved in the action of a hormone called fibroblast growth factor. The result is the condition called achondroplasia, in which the limb bones fail to elongate normally so that the affected person has short stature. In most case the parents do not have this condition, so it is the result of a new mutation. In about 98% of cases, the mutation is a single base change that results in the replacement of one amino acid by a different amino acid in the protein that binds to the growth hormone.

On the other hand, a very small fraction of mutations are positive in their effects – they increase evolutionary fitness. An example of a positive mutation that has been selected for recently in human history is a mutation that occurred in the gene encoding the enzyme lactase. This enzyme is required so that babies can digest the sugar lactose that they receive in their mother's milk. After weaning, the production of this enzymes ceases, as babies prepare for an adult diet. Several thousand years ago, a mutation occurred in a human that allowed him or her to continue to digest lactose in dairy products into adulthood. From the ethnic distribution of this trait, it is likely that this mutation occurred in a European who lived in an area where animals such as cows and goats had been domesticated, and thus where a supply of milk was available to adults. This mutation spread throughout Northern Europe by positive natural selection because it conferred survival and reproductive advantages on those who carried it. Today this mutation occurs in 95–98% of Europeans, but in only 20–50% of Hispanics and 5% of Asians. Adult people who lack this mutation suffer from the condition called "lactose intolerance" because the inability to digest lactose results in bloating and cramping.

It is the few mutations that have positive effects that provide the variation whose selection drives the evolutionary process. So we come to the surprising realisation that the diversity of life is created by mistakes – errors made when DNA is copied. If mutations did not occur, evolution would not be possible. *Evolution is the result of a series of successful mistakes.*

So the conclusion that biologists have reached is that, because both mutation and natural selection are observable facts, evolution is inevitable. *Thus evolution is both a theory and a fact.* But evolution is more than just another scientific theory because it challenges those views that suggest humans are basically different from other animals and so can escape the laws of nature. It is this aspect of evolution that makes it so unattractive to many people. But rejecting evolution means that we reject the best means we have found so far to understand ourselves and our place in the world.

Hierarchical Classification

We humans have a strong desire to order objects into categories because this helps us to make sense of the world. Before experimental biology got underway at the end of the nineteenth century, most biologists were occupied in trying to classify all the living organisms they could find into groups. The characters they used for this purpose were morphological ones, but today we have, in addition, the sequences of DNA and proteins. All organisms can be classified into groups, using the system promoted by the Swedish biologist Carolus Linnaeus in the eighteenth century. Linnaeus used five groups but today this has expanded to eight groups, called domains, kingdoms, classes, orders, phyla, families, genera and species. These groups are defined by agreed sets of characteristics. For example, the kingdom of metazoa, or animals, is defined as those multicellular eukaryotes that eat other organisms, not including organisms such as *Darlingtonia* that are obviously plants. It is important to appreciate that of these eight groups, only the species has a real existence – all the other groups are constructs of the human imagination that selects certain characters as being more useful for classification purposes than others. Species can be said to have an existence independent of how humans view them because the most commonly used definition of a species is a group of interbreeding individuals.

Each of these eight groups is defined by an agreed set of similar characteristics, but the important feature of this scheme for the point of view of evolutionary theory is that these sets are nested within one another. This nested pattern of groups

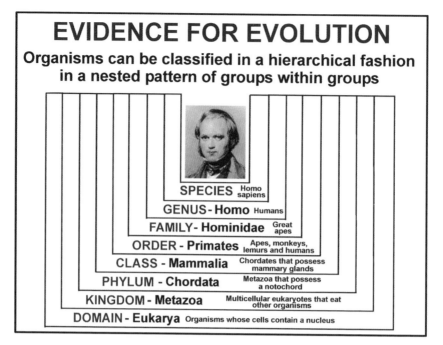

Fig. 4.21

within groups is consistent among many different traits, from the anatomical to the molecular, with each group defined by a unique set of characters. As we move up the hierarchy, the organisms within the groups become less and less similar. This hierarchical classification is explained by Darwin's idea of "descent with modification" because this idea predicts that groups of organisms are similar because they have a common ancestry. The application of this hierarchical scheme to the human species is illustrated in Fig. 4.21, which contains short definitions of each group.

Biogeography

Biogeography is the study of biodiversity across all regions of the Earth. It aims to determine not only the patterns of distribution and abundance of each species but also what factors determine these patterns. These factors include such historical events as continental drift, glaciation, and extinction, but of especial importance is where each species originated. You might think that a given species should be found wherever a suitable climate and food source are available, but it turns out that geographical location is a better predictor of where similar species live than either of these factors. Thus similar climatic regions contain very different animals. The two major scientists who first came to this realisation were Wallace and Darwin, and they both interpreted this finding in the same way – biogeography reflects ancestry rather than climate, because each species has evolved from other species at a particular location, and then migrated outwards from the point of origin until a barrier was reached.

The father of biogeography was Alfred Russel Wallace (1823–1919), whose co-discovery of natural selection, you will recall, prompted Charles Darwin to publish *On the Origin of Species*. Wallace was the leading expert on the distribution of animal species in the nineteenth century, and worked in both the Amazon River basin and the Malay Archipelago. He collected more than 125,000 specimens in the Malay Archipelago, of which about 80,000 were beetles – I pointed out in the section on biodiversity that most animals are insects, and that most insects are beetles. His name is commemorated in the term "the Wallace line", which describes the fact that there is a clear separation of land species between the southeastern and northwestern parts of Indonesia. The Wallace line is indicated in Fig. 4.22.

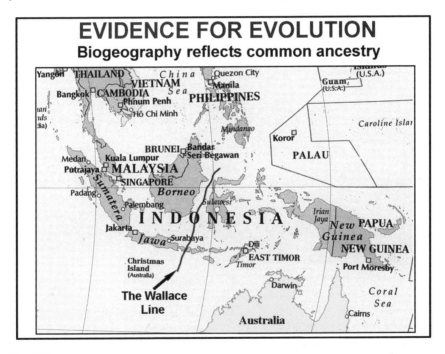

Fig. 4.22

The land animals and plants to the northwest of the Wallace line are very similar to those found in other parts of Asia, but those to the southeastern side are more similar to those found in Australia. This separation seems at first sight to be arbitrary, but today we know from the geological evidence of past sea levels that the Wallace line marks an ancient deep-water passage that separated the two land masses, even when the sea levels were one hundred metres lower than they are today. This deep water presents a barrier to the passage of most land animals and plants. This distribution pattern is explained by evolution if you assume that each species appears at a particular location, and then gradually disperses away from its point of origin until it meets a barrier that it is unable to cross.

Darwin devoted two chapters in his famous book to the geographical distribution of species. The first sentence reads:

> In considering the distribution of organic beings over the face of the globe, the first great fact that strikes us is, neither the similarity nor the dissimilarity of the inhabitants of various regions can be wholly accounted for by climatal and other physical conditions.

As an example, Darwin points out that both the North and South American continents share the same range of humid forests, arid mountains, grassy plains, marshes, lakes and rivers as does the European continent, but that there are almost no species of animals and plants that occur in both locations. A further example is the similarity in climate between parts of Australia, South Africa and western Southern America. But this similarity is not accompanied by any similarity in the animals and plants that occur in these regions. Kangaroos were not found in Europe and rabbits were not found in Australia, until humans transported them there.

Similar observations have been made for marine environments. The shelled animals such as crustaceans and sea urchins, are quite different between the eastern and western shores of South America. On the other hand, many of the fish species are the same on the opposite sides of the isthmus of Panama, suggesting that in the recent geological past, there was a free flow of water between the Pacific and Atlantic oceans. Since Darwin's time, geologists have established that the isthmus of Panama formed about three million years ago, an event that had dramatic effects on the world's climate. The blockage of exchange between the two oceans resulted in a rerouting of ocean currents, the most important being the formation of the Gulf Stream that today keeps Britain and northwestern Europe at a habitable temperature in the winter. The formation of the Panama land bridge allowed animals and plants to migrate between North and South America, an event called by fossil experts the "Great American Interchange". For example, the ancestors of the opossums, armadillos and porcupine found in North America today came across the newly-formed land bridge from South America, while the ancestors of animals such as cats, bears and racoons went in the opposite direction.

Another observation that impressed Darwin is that, although species at different locations on a continent are distinct, nevertheless they are more similar to one another than to those on other continents. He pointed out that if you travel from north to south on the same land mass, successive groups of species replace one another but are also closely related. The same conclusion applies to islands. Their inhabitants are distinct from those on the nearby mainland, but are more closely related to them than to those on different continents. During his stay on the Galapagos Islands, Darwin collected a number of different mockingbirds as shot specimens. On his journey home he realised that all the mockingbirds caught on each island were of the same species, but different from those caught on nearby islands, while all the island mockingbirds were similar to those he had found in Chile. It is this observation that made him wonder about the popular idea that he had previously accepted – that species are stable once created and do not change. He speculated that this distribution could be explained by another hypothesis – that the islands were originally populated by a few individual birds that had managed to survive the journey from the mainland and, finding themselves on different islands with no competing species, had changed over time into new species characteristic of each island. It was this type of observation that sparked in Darwin's mind the train of thought that led eventually to the idea of natural selection as the mechanism of change.

Vestigial Structures

In the penultimate chapter of *On the Origin of Species*, Darwin discusses what he terms "rudimentary structures" as one of the lines of evidence for evolution. Today we call these "vestigial structures", and they are defined as structures that are reduced in complexity and function compared to similar structures in other organisms. These structures occur at both the anatomical level and at the molecular level.

A common misconception is that vestigial structures are necessarily functionless, but this is not an essential part of their definition. For example, the wings of flightless birds such as ostriches and emus are vestigial structures. This does not necessarily mean that these structures are functionless – ostriches use their rudimentary wings to shade their chicks, for insulation, for courtship display and as sails when running across the plain. The point is that it is clear from studying other organisms that wings are complex adaptations that enable powered flight, but that they are too small in ostriches to do this. This apparent paradox can be explained in evolutionary terms if you assume that ostriches evolved from ancestors that did fly, but who then adapted to life on the plains by abandoning flight to avoid predators in favour of increased body size, swift running and large feet that pack a powerful kick. In the same fashion, some genes have mutated so that they have lost their original functions and are called pseudogenes. Some pseudogenes have acquired different functions unrelated to their original functions. Natural selection works essentially as a tinkerer rather than a designer. It fiddles with what it has got to meet the immediate needs imposed by the environment, rather than progressing to a predetermined end point.

There are many examples of vestigial anatomical structures that have no known functions, such as reduced pelvises and hind limb bones inside the skins of pythons and whales, remnants of eyes in fish and salamanders that live inside caves and mole rats that live underground, flightless beetles that have perfectly formed wings uselessly located inside fused wing covers, and wisdom teeth, appendix and coccyx in humans. Figure 4.23 illustrates some vestigial structures.

The ape ancestors of humans were herbivorous and used molar teeth to chew and grind tough plant material, which was digested in a large sac in the digestive tract called the caecum. The word "caecum" is the Latin for "blind" and refers to

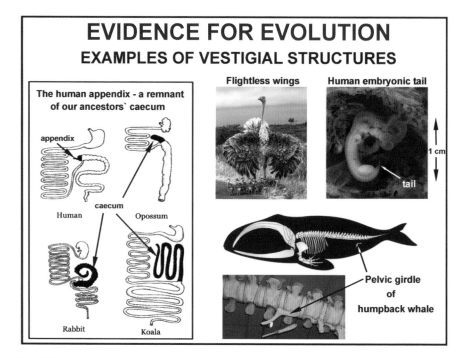

Fig. 4.23

the fact that the bottom of the caecum is a dead-end – it is a side pouch in the digestive tract. The caecum is essential for the digestion of plant material because it contains bacteria that can break down cellulose, which animal cells cannot. Because the caecum is a dead-end sac, it allows a permanent colony of cellulose-digesting bacteria more time to digest tough plant material. The size of the caecum in different animals reflects the proportion of the diet that is derived from plants (see Fig. 4.23).

Modern humans are omnivorous and so lack a caecum, but in its place is a smaller organ called the appendix, which is essentially the remnant of the apex of the caecum that our ancestors possessed. The appendix is not essential for life and indeed may need to be surgically removed if it becomes infected. It may have some functional roles, such as providing a reservoir of bacteria that aid digestion.

On the right of Fig. 4.23 is shown an example of vestigial pelvic girdles in whales and the occurrence of a tail in early human embryos. All the whales and dolphins have vestigial pelvic girdles, and many have, in addition, the remnants of femurs. These structures no longer function in locomotion, and are not attached to the vertebral column, as they are in fossil ancestors of whales. All mammals have a tail at some point in their development, whether or not the adult animals have tails. In humans, the tail is present during the first one to two months after conception, and is then destroyed by a destruction pathway called programmed cell death. Figure 4.23

shows a five week-old ectopic human embryo with a clearly visible tail arrowed. The tail is located at the end of the coccyx, a series of four fused vertebrae at the base of the spine that persists in the adult human. The coccyx is the remnant of the embryonic tail that is left after the tail has been destroyed. There are rare reports of some human babies being born with tails that can be up five inches in length. Such tails result because the destruction of the embryonic tail fails to go to completion, probably because of mutations in the developmental or programmed cell death pathways. The genes that control tail development in mice have been identified. Variant mice that do not develop tails have a mutation in one of these genes that reduces its activity. So a plausible explanation of the rare examples of human tails is that another mutation has increased the activity of this gene in these individuals. A similar explanation would account for occasional reports of whales with external hind legs.

Vestigial structures at the molecular level are called pseudogenes. These are defined as genes that have lost their original protein-coding ability. They can be identified because enough of the original gene sequence remains to enable its original function to be recognised by comparison with similar genes that have retained their function. About 19,000 pseudogenes have been identified so far in the human genome. This number is almost the same as the number of protein-encoding genes so far identified, which is 19,042. Good examples of pseudogenes are found in the large family of genes encoding the proteins that enable animals to smell.

Most animals possess cells that enable them to detect chemicals in the environment. In mammals the sense of smell is created by membrane proteins located at surface of cells lining the nasal passages. These proteins are called olfactory receptors because they are capable of binding a very large range of different chemicals present in the air. Each protein detects only one type of such chemical, so there are many such proteins. They can be recognised by two features; they all fall into one class of proteins with very similar amino acid sequences, and hence similar conformations, and they are found only in the cells lining the nasal passages. These cells are called olfactory neurons because they are nerve cells, whose other end is located in the olfactory centre in the brain where the signals coming along the nerve fibres are interpreted as distinct smells. There are more than one thousand olfactory receptor genes in the human genome, making it the largest family of related genes in this species – a superfamily. But more striking still, is that about 60% of these genes occur as pseudogenes in humans – that is, their sequence has mutated in such a way that the complete proteins can no longer be made.

How can we account for the occurrence of some many pseudogenes in this human superfamily? The suggested explanation is that the animal ancestors of humans relied on their sense of smell to a greater extent than humans do. In support of this evolutionary explanation is the finding that in mice only 20% of this superfamily occurs as pseudogenes, while in chimpanzees and gorillas the fraction of pseudogenes is around 30%. As human ancestors became more reliant on vision than on smell, mutations inactivating the olfactory receptor genes became less susceptible to natural selection, and so they persist today in the form of pseudogenes. It is possible that some of these pseudogenes have acquired new functions, because although they no longer encode proteins, many of them are still transcribed into RNA, A recent discovery is that some pseudogenes that do not encode proteins still produce functional products, in the form of RNA molecules that regulate the activity of the original gene from which they are derived. So some pseudogenes may not dead-end products of mutation, but form a potential source of new functions if natural selection happens to favour the effects of further mutations in these genes.

The widespread occurrence of vestigial structures is predicted by evolutionary theory. If it is correct that all organisms are descended from a common ancestor, then both structures and functions necessarily will have been gained and lost as new species arise. In the section entitled "Facts, Theories and Hypotheses" in Chapter 2, I pointed out that to be considered scientific, a hypothesis must be capable of being falsified. This requirement is admirably met by vestigial structures. According to evolutionary theory, no organism can have a vestigial structure that was not previously functional in its ancestors. It follows that if evolution is an incorrect description of the living world, we could find vestigial structures that lack an evolutionary explanation. For example, evolution would be falsified if we found vestigial feathers or wings in mammals, apes with vestigial gizzards, vestigial spinal columns in crustacea, or pseudogenes encoding the enzymes of cellulose biosynthesis in mammals. So far, evolutionary theory has withstood such potential falsification.

Further Reading

1. *Evolution:* Nicholas H. Barton, Derek E.G. Briggs, Jonathan A. Eisen, David B. Goldstein and Nipam H. Patel. Published by Cold Spring Harbor Laboratory Press, 2007. ISBN 978-0-87969-684-9.
2. *Evolutionary Analysis:* Scott Freeman and Jon C. Herron. Published by Pearson Education, Inc. 2007. 4th edition. ISBN 0-13-239789-7.
3. *Evolution:* Mark Ridley. Published by Blackwell Publishing, 2004, 3rd edition. ISBN 1-4051-0345-0.
4. *Understanding Evolution: History, Theory, Evidence and Implications.* Geoff Price, 2006. www.rationalrevolution.net/articles/ – scroll down to 'Understanding Evolution: History, Theory, Evidence, Implications'. This is a comprehensive account of the historical relations between science and religion in the context of evolution.
5. *The Making of The Fittest: DNA and the Ultimate Forensic Record of Evolution.* Sean B. Carroll. Published by Quercus, 2006. ISBN 978 1 84724 4765. This book is a very readable account for the non-specialist of the evidence for evolution that resides in the sequences of DNA molecules.
6. *Your Inner Fish.* Neil Shubin. Published by Allen Lane, 2008. ISBN 978-0-713-99935-8. A very readable account of the evidence that many features of the human body evolved from those of animals that once swam in the oceans.
7. *The Real life of Pseudogenes.* Mark Gerstein and Deyou Zheng. Scientific American, 295:22 49–55, August 2006. A recent account of research on pseudogenes that emphasises that some have acquired new functions.

Chapter 5
The Evolution of Eyes

One of Darwin's Difficulties

One of the many impressive aspects of *On the Origin of Species* is that Darwin devotes an entire chapter to what he terms "Difficulties of the Theory". This is not a habit common among leading scientists today! One of the difficulties he discusses is the evolution of eyes. He starts by pointing out what many people feel when considering this organ in terms of evolutionary theory:

> To suppose that the eye with all its inimitable contrivances for adjusting the focus to different distances, for admitting different amounts of light, and for the correction of spherical and chromatic aberration, could have been formed by natural selection seems, I freely confess, absurd in the highest degree.

The human eye is superbly adapted to its function of providing detailed visual information about the world. Figure 5.1 shows a section through the human eye, to remind you of its basic features.

The eye is roughly spherical, located in a protective bony socket and surrounded by six muscles. Unlike a camera, the eye forms a sharp image of only a small part of the visual field at any one time, so these muscles are in constant use to allow the eye to scan a wide angle, so that the brain can build up a detailed image of the whole scene. Light enters through a transparent protective cover called the cornea, whose surface is kept clean by means of secretions from tear glands located above and at the inner side of the eye. These secretions contain antibacterial proteins. The light then passes through an opening in the iris called the pupil; the diameter of this opening is changed automatically to accommodate different light intensities by means of muscle fibres located in the iris. The iris is coloured to prevent light entering except through the pupil. Light continues through the lens and is focussed onto the light-sensitive layer called the retina that lines the inside of the eye. The lens is made of transparent cells arranged in a biconvex shape. This shape can be changed to focus on objects at different distances by means of muscles that surround the lens.

The retina consist of an outer pigmented layer that functions to prevent light reflecting inside the eye, and an inner layer of nerve cells, some of which are sensitive to light (see Figure 5.1, right-hand diagram). It is this pigmented layer that

J. Ellis, *How Science Works: Evolution*, DOI 10.1007/978-90-481-3183-9_5,
© Springer Science+Business Media B.V. 2010

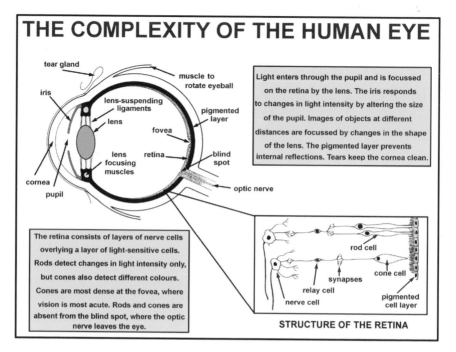

THE COMPLEXITY OF THE HUMAN EYE

Light enters through the pupil and is focussed on the retina by the lens. The iris responds to changes in light intensity by altering the size of the pupil. Images of objects at different distances are focussed by changes in the shape of the lens. The pigmented layer prevents internal reflections. Tears keep the cornea clean.

The retina consists of layers of nerve cells overlying a layer of light-sensitive cells. Rods detect changes in light intensity only, but cones also detect different colours. Cones are most dense at the fovea, where vision is most acute. Rods and cones are absent from the blind spot, where the optic nerve leaves the eye.

STRUCTURE OF THE RETINA

Fig. 5.1

causes the pupil to look black. Notice that the light-sensitive cells receive light only after it has traversed the covering layer of nerve cells that send impulses generated by the light to the brain via the optic nerve. This does not seem to be the optimum position for the light-sensitive cells, so the human eye may not be as perfect as Darwin thought – an engineer designing the most efficient camera puts the film directly in the path of the light. By contrast, the eyes of molluscs such as the octopus have a better design – in these species, the light-sensitive cells form the top layer of the retina.

There are two types of light-sensitive cells, called rods and cones after their shapes. Rods and cones are nerve cells specialised to absorb photons of light, whose absorption triggers electrical impulses that are relayed by synapses to neighbouring nerve cells, and so onto the brain. Rods detect light intensity only, and so enable black and white images to be formed in the brain, but the cones in addition detect different wavelengths of light and so enable colour vision. The rods are more sensitive to light than the cones, which is why colours are so difficult to see in dim lighting. But the cones can resolve more detail than the rods. The reason for this difference in resolving power between rods and cones is that only one cone cell is connected to the next nerve cell in the relay, but several rod cells are connected to the next nerve relay cell (see diagram of the retina in Figure 5.1). There are about one hundred million rod cells in the human eye, but only about three million cone cells. The densest concentration of cone cells occurs at a point called the fovea, which is directly on the optical axis of the lens and provides the sharpest image.

A rod cell generates a nerve impulse after absorbing just one photon of light, although the brain requires impulses from six photons to perceive the signal. This remarkable sensitivity is mediated by a light-absorbing molecule, a protein-pigment complex called rhodopsin. This complex consists of a membrane protein called opsin, bound covalently to a pigment called retinal that absorbs photons. Retinal is made in the retina from vitamin A, so that a diet deficient in this vitamin leads to a condition called night-blindness, in which the patient experiences difficulty in seeing in dim light. Absorption of a photon of light by rhodopsin results in a very rapid and reversible change in the conformation of the retinal that triggers conformational changes in other proteins bound to the retinal membrane. These changes in protein shape generate an electrical impulse that travels down the axon of the rod cell to the next nerve cell in the relay.

Cone cells contain three slightly different kinds of opsin protein, the differences being in a few of the amino acids in the polypeptide chain, close to where the retinal is bound. These changes in some amino acids result in different absorption spectra of the attached retinal pigment because they alter the electronic distribution in the retinal molecule. The absorption maxima of these three rhodopsins are at 560, 530 and 426 nm, resulting in cones sensitive to red, green or blue wavelengths of light. Each cone contains only one type of these three rhodopsins, an arrangement resembling that found in the photosensors of digital cameras. The rhodopsin in rod cells has its absorption maximum at 500 nm, which is the most abundant wavelength in solar radiation at ground level.

Many of the details of the "inimitable contrivances" that allow the eye to function so well were discovered after Darwin's time, but he know enough to appreciate the problem of explaining its origin in term of natural selection. After pointing out how well the eye is adapted to its function, Darwin goes on to say in his chapter "Difficulties with the Theory":

> Reason tells me, that if numerous gradations from a simple and imperfect eye to one complex and perfect can be shown to exist, each grade being useful to its possessor, as is certainly the case; if further, the eye varies and the variations are inherited, as is likewise certainly the case; and if such variations be useful to any animal under changing conditions of life, then the difficulty of believing that a perfect and complex eye could be formed by natural selection, though insuperable to our imagination, should not be considered as subversive of the theory.

In the reminder of his discussion about the evolution of eyes, Darwin points out that much simpler eyes than the ones than mammals possess occur in animals such as starfish and lancelets. These "eyes" consist only of some pigmented cells shielding some photosensitive cells on one side They can sense the direction of light, but are unable to form images because they lacks a lens, and so are more accurately called "eyespots". He suggests that the existence today of these eyespots supports his idea that "simple and imperfect eyes" could be ancestors of modern eyes. Since his time, many more examples of biological light-detection structures have been found, and I discuss some of these in the next section.

Light-Detection Structures

The biological world contains an amazing range of different types of light-detecting structure. It is convenient to divide them into two broad categories, termed eyespots and eyes. These categories are not sharply separated but grade into each other, as would be expected if the more complex structures have evolved from simpler structures. Eyespots are defined as structures that detect the intensity and direction of light, and so allow organisms to move towards or away from the light source, but are unable to determine the light intensity from more than one direction at the same time, so they cannot form an image. Eyes are defined as structures that can form images, often by means of a lens or mirror, but not necessarily so – some eyes form images using a small hole, in the manner of a pinhole camera.

About two-thirds of the thirty-three metazoan phyla possess light-detecting structures, but only six of these phyla possess eyes that form images. But these six phyla contain about 96% of all known animal species alive today. These numbers indicate the high value of possessing eyes. The first animal that developed eyes that enabled it to catch prey and avoid predators would have a huge advantage over its competitors. So the appearance of eyes is thought to have triggered an arms race, in which competing animals were forced to evolve better and better eyes, as did their prey and predators. The first eyes are suggested to have appeared before 540 million years ago, when there is a sudden appearance of many animal phyla in the fossil record, an event termed the "Cambrian explosion". The word "Cambrian" is derived from the Latin for the country of Wales, where the first evidence of this increase in animal fossils was discovered. One reason for this explosion may be that the arms race triggered by the appearance of eyes favoured the evolution of hard shells for protection; such shelled animals fossilize much more readily than the soft-bodied animals that preceded them.

It is plausible to think that before eyes appeared there were only eyespots, that enabled organisms to move towards or away from the light, depending on their life style. So we may gain clues about the likely series of evolutionary events by comparing the structures and functions of eyespots and eyes in modern organisms. Figure 5.2 illustrates some examples of eyespots. The simplest eyespot known consists of a cluster of membrane-bound protein-pigment molecules found inside unicellular eukaryotes such as *Euglena* and *Chlamydomonas*. These organisms are photosynthetic and so need to find an illuminated environment in order to survive. The absorption of light by the eyespots triggers a metabolic pathway that controls the rotation of the flagellum. In *Euglena* the eyespot lies in the cytosol near the flagellum, but in *Chlamydomonas* it lies inside the single chloroplast. Both the protein and the pigment in the eyespot of *Euglena* are different from those in *Chlamydomonas*, suggesting that they evolved independently.

One way to improve the simple eyespot to make it better able to determine the light direction would be to shield it from one side by means of light-absorbing pigments, so that light can fall on the light-sensitive protein-pigment complex only from the other side. Precisely this arrangement is found in the larva of the box jellyfish, *Tripedalia*, where a cup of pigment granules embraces the membranes containing the light-sensitive protein. Interestingly, the adult box jellyfish also possesses such simple eyespots but in addition has true eyes, complete with lens, iris and retina. The box jellyfish does not contain a brain, which tells us that a brain is not necessary for eyes to evolve. It is common for individual invertebrate animals, especially annelids and molluscs, to possess more than one type of light-detecting structure.

A further elaboration of the simple eyespot would be for the shielding pigment granules to occur in an adjacent cell to the one containing the light-sensitive protein-pigment; this arrangement occurs in the larvae of the ragworm, *Platynereis*. If the number of light-sensitive cells shielded by the adjacent cells then increases, the resulting structure can determine the light intensity from more than one direction at the same time – this type of structure is found in the flatworms called planaria. Planarians seek the shade, so use their eyespots to find dark environments where they are safer from predators. All these stages are illustrated in Figure 5.2.

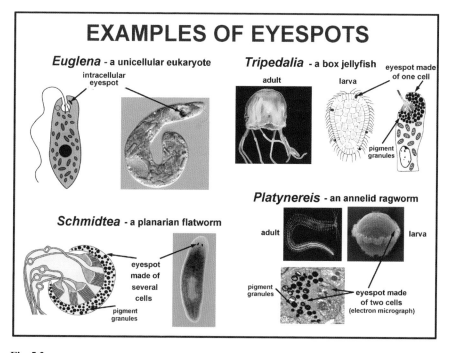

Fig. 5.2

The next step from a planarian eyespot would be for the structure to deepen further into a pit and contain more light-sensitive cells, so that the direction of light can be determined more precisely. Pit eyes are common in invertebrates, such as the mollusc *Nautilus* (see Figure 5.3). The diameter of the opening in the *Nautilus* eye can be varied by a factor of seven-fold to accommodate varying intensities of light. There are now two evolutionary possibilities: either the pit eye can develop into a larger structure containing more light-sensitive cells called a chambered eye, or the entire pit eye can be duplicated many times to form a compound eye. The essential distinction between these two forms is that there is only one optical system in chambered eyes, but mutiple optical systems in the compound eye. The most detailed images are formed by chambered eyes because the amount of detail possible in the pit eye is limited by the diameter of the opening – the smaller the opening the better the detail, but the less the amount of light that falls on the light-sensitive cells. In chambered eyes this problem is overcome by the evolution of the lens, which enables detailed images to be formed at low light intensities.

Chambered eyes are sometimes called simple eyes, but this is a poor name because they are far from simple! The pit eye is initially open at the top to admit light, but if this opening is covered by a transparent layer of cells to prevent the pit being blocked by detritus or parasites, it allows the pit to be filled with transparent liquid, which improves its optical properties. This in turn, allows the subsequent development of internal devices that dramatically improve the quality of the image, such as an iris, lens or mirror. The octopus has a lens-containing eye very similar to the human eye, while the scallop *Pecten* has a mirror eye (Fig. 5.3). The lens is made of concentric layers of highly elongated cells packed with transparent proteins called crystallins. Analyses of the amino sequences of crystallins reveal that they are closely related to ordinary proteins involved in metabolic pathways – they have been co-opted for vision because they happen to have the right degree of

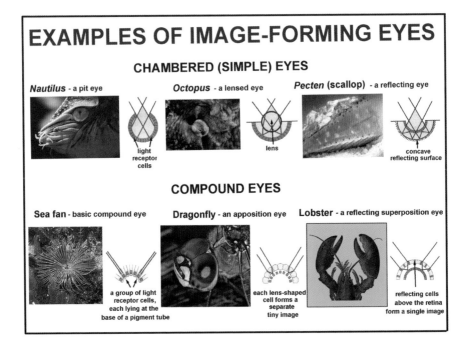

Fig. 5.3

transparency when present at high enough concentration to refract light rays This is another example of a common theme in evolution – one thing leads to another. Thus a structure can be used for a quite different function from the one for which it initially evolved – in the section on transitional forms I mentioned how the appearance of feathers for insulation allowed the subsequent development of flight. In the mirror eye, a concave inner layer of cells reflects light back onto the layer of light-sensitive cells. The reflecting material is made of crystals of the base guanine – one of the four bases found in DNA. You will recall from your knowledge of physics that there are two ways of focussing light used by the manufacturers of microscopes and telescopes – either a convex lens or a concave mirror. Evolution discovered both these ways long before humans appeared. Figure 5.3 illustrates some image-forming eyes.

In the simplest compound eye each light-sensitive cell lies at the base of a tube of pigment that restrict the light to a narrow angle, that can be as small as two degrees. Figure 5.3 shows this type of eye found in the sea fans, a group of sessile marine invertebrates related to jellyfish. A more sophisticated version of the compound eye is called the apposition eye, common in insects such as the dragonfly. Apposition means "standing side by side" and refers to the multiplicity of identical light-sensitive units that are not in contact, as they are in the retina of chambered eyes. Each unit may have its own lens and so forms a tiny image. There are also superposition eyes, found in some nocturnal insects and deepwater crustaceans, where many individual lenses co-operate to form a single image. There are even superposition eyes that use reflecting layers rather than lenses – Figure 5.3 shows the example of the lobster eye.

Plausible Evolutionary Possibilities

Eyes being soft-bodied structures, it is not surprising that the fossil record of extinct eyes is so poor. But what we can do is to see if we can arrange the existing eye structures into a plausible pathway. The term "plausible" here means that the pathway must include existing examples, and that we must be able to specify the advantage of each stage over its presumed precursor – a requirement of evolution by natural selection.

Figure 5.4 illustrates the main stages in a plausible pathway.

We start with an eyespot that expands into a flat patch of light-sensitive cells, linked to one or more neurons that can generate an electrical impulse when light falls on the patch. This simple arrangement enables the organism to detect light intensity but not direction, and this is sufficient to enable it to move towards or away from the light source. An improvement to this system would be for the patch to be shaded on one side by cells containing light-blocking pigment granules. If this shaded patch then deepens into a pit, the intensity of light can be measured in different directions at the same time. If the chamber gets larger, more light-sensitive cells are located together, creating a retina. This chamber will initially be filled with

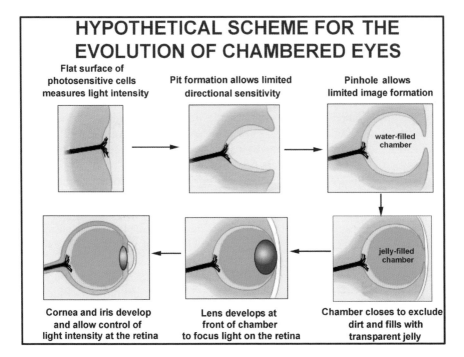

HYPOTHETICAL SCHEME FOR THE EVOLUTION OF CHAMBERED EYES

Flat surface of photosensitive cells measures light intensity

Pit formation allows limited directional sensitivity

Pinhole allows limited image formation

water-filled chamber

jelly-filled chamber

Cornea and iris develop and allow control of light intensity at the retina

Lens develops at front of chamber to focus light on the retina

Chamber closes to exclude dirt and fills with transparent jelly

Fig. 5.4

water, creating the danger of clogging by detritus and parasitic organisms. These dangers are averted if the pit opening becomes covered by a transparent layer of skin. If the aperture of the pit reduces in diameter, the resolution of light direction improves, but the amount of light that is admitted is also reduced, which would be a disadvantage in low-light environments.

This problem is solved if a lens forms in the transparent jelly that now occupies the chamber. This could result from the splitting of the transparent layer over the eye opening into two layers. The outer layer forms the cornea and the inner layer the lens. The formation of a lens creates two large advantages – more light can be admitted, and it can be focussed onto the retina so that a detailed image is possible. The formation of a cornea both protects the lens and assists in refracting light rays, while the formation of an expandable iris permits the eye to operate under a wide range of light intensities.

How long might it take for such a pathway to evolve, and is there any way we can observe some of the changes in real time today? Biologists have constructed a model for the evolution of the chambered eye that attempts to answer these questions. This model assumes that each mutation causes a change of 1% in some character relevant to eye formation – this might be the size of the light-sensitive patch or the concentration of a particular protein. On this basis, the number of sequential mutations required to form a chambered eye from a light-sensitive patch is 1829. Applying population genetic equations to this number of mutations gives a figure of nearly

364,000 generations of the organism for the chambered eye to develop by natural selection. If we now suppose that each generation takes one year, which is a reasonable time for a small aquatic animal, the eye could evolve in less than half a million years. This compares with the minimal estimate of thirty million years between the appearance of the first simple animals in the fossil record and the Cambrian explosion some 540 million years ago. This calculation implies that about 500 million years ago, the early vertebrates possessed eyes basically similar to our own.

Could we observe any of these small incremental changes in the laboratory today? On the model described above, it takes about two hundred mutations to change a flat surface into a slightly curved surface. If the generation time of the organism is one year, it would need an experiment lasting about 36,000 years to observe this change. Even if the generation time of the organism was one day, it would still take about one hundred years. This calculation illustrates the basic problem with trying to convey the nature of evolutionary change to the person-in-the-street – the rate of change is so slow as to be beyond human comprehension.

An unsolved problem is how the first rhodopsin protein appeared and what function it served. There are proteins called rhodopsins in Bacteria and Archaea, involved in using light energy to move ions across the cell membrane, but this name is misleading. These prokaryotic proteins are called rhodopsins because they have a similar conformation to the eukaryotic rhodopsins, but there is no discernable similarity in amino acid sequence between these proteins and the eukaryotic rhodopsins. On the other hand, there are many Bacteria and Archaea that have not yet been studied, so it is possible that the origin of the eukaryotic rhodopsins will be found among them.

Genetic Control of Eye Formation

In the 1970s so many different types of eye were known that it was proposed that the eye has evolved independently at least forty times in different lineages of animals. At that time, the available information was largely derived from studies of structures visible in the light and electron microscopes. This picture was challenged by the discovery in the 1990s that a single developmental regulatory gene, called P*ax6, is* required for eye development in species as different as insects, mice and humans. There is now much more information available about the genetic control of eye formation in a range of invertebrate and vertebrate species, leading to the proposal that some common genetic factors underlie the evolution of different types of eye. This conclusion reinforces the general conclusion that I have emphasised before in this book – all organisms are related to one another no matter how different they appear to us. The *Pax6* gene links us to our early animal ancestors.

The first gene controlling eye formation was discovered in the fruit fly *Drosophila* in 1915. This gene was called *eyeless* because the result of a mutation in this gene is to prevent the formation of eyes. It is common practice to name a gene in

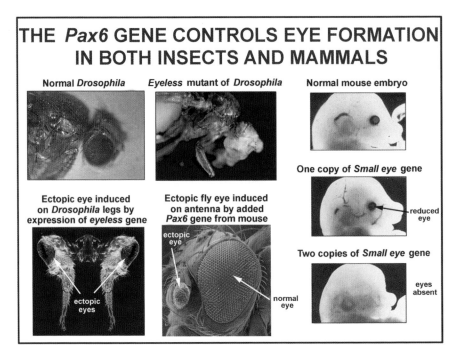

Fig. 5.5

terms of the effect produced by a mutation in that gene; this name is written by convention in *italics*. Figure 5.5 compares the head of an *eyeless* mutant of *Drosophila* with that of the normal, wild type fly. It was seventy-six years later before a similar mutation was found in mice, and this gene was called *Small eye*. Mice that contain one copy of the mutant gene have reduced eyes, but those with two copies lack eyes entirely, as well as the nose and a large part of the forebrain, and die before birth. The right-hand part of Figure 5.5 shows the effect of this mutation in mice embryos. This mutation in mice resembles a rare human genetic condition called aniridia, where one copy of the mutant gene causes reduced irises, but two copies result in no eyes or nose, and early death. The key connection between these genes was made in the 1990s when it was discovered that the amino acid sequence encoded by the *eyeless*, *Small eye* and *aniridia* genes are very similar to one another and to a similar gene in zebrafish, amphibians, sea urchins, squid and planarians.

This was a surprising finding because until that time the very different eye structures found in vertebrates and invertebrates had supported the idea that they had independent origins. This discovery prompted the different view that *Pax6* is a master regulatory gene involved in determining eye development in most, if not all, animals, and thus that there was some common ancestry in the early evolution of different types of eye. This change of view is another example of the provisional nature of all scientific ideas. You will recall that I pointed out in Chapter 2 that all scientific ideas are provisional because they may be changed in the future as the result of new discoveries.

Some of the experimental evidence in support of this new view is shown in the lower left-hand side of Figure 5.5. Genetic engineering techniques allow scientists to insert genes into the eggs of *Drosophila* so that they are expressed in tissues where they are normally switched off. If the *eyeless* gene is expressed in the cells that give rise to legs, eyes develop on these legs. Structures that appear in an abnormal location are called "ectopic". Ectopic eyes have also been made to appear on antennae and wings of *Drosophila* by targeting the expression of the inserted *eyeless* gene to the precursors of these organs. But a more amazing discovery is that the *Pax6* gene from the mouse will also cause ectopic eyes to appear on the antenna of *Drosophila*. This eye is a typical fly eye, not a mouse eye, because it requires over 2000 other genes to make an eye and in this experiment only one of these genes, the *Pax6* gene, is derived from a mouse.

How is the *Pax6* gene thought to work? This gene encodes a protein that acts as a transcription factor. You will recall from Chapter 4 that transcription factors are proteins that bind to regions of DNA that control the expression of other genes. The binding of a transcription protein onto a particular sequence of bases in the DNA either switches the gene(s) on or switches it off. The latter gene(s) may also encode another transcription factor(s) and so regulates the expression of yet more genes. In this way, a hierarchical cascade of events can be initiated by one master regulatory gene. This mechanism of gene regulation was discovered in the 1950s by French scientists working on how bacteria adapt their metabolism to utilise different nutrients they encounter in their environment, but it is now clearly established that the same principle underlies the development of specialised organs in the most complex organisms on the planet. *All organisms are related to one another, from bacteria to humans.*

Further Reading

1. *Animal Eyes*. Michael F. Land and Dan-Eric Nilsson. Oxford Animal Biology Series. Published by Oxford University Press, 2002. ISBN 0-19-850 9685.
2. *Master Control Genes in Development and Evolution:* Walter J. Gehring, Yale University Press, 1998. ISBN 9780300074093.

Concluding Remarks

The Greek philosopher Socrates famously wrote that the unconsidered life is hardly worth living. What he meant by this remark is that the well-lived life is one that has goals and principles that are chosen by the one who lives it, rather than imposed by others. We are all born into a particular set of historical circumstances that may limit the possibilities open to us, but a considered life is always enhanced by thinking about things that matter, such as our aims, our values, and how best to cope with the problems we will encounter. Integral to the considered life is to come to some rational view about the nature of the world and our place in it.

In this book I have tried to introduce some clarity into two areas that are often misunderstood and misinterpreted: the nature of science and the science of nature. It is important to understand the nature of science because its application has created and sustains the modern developed world. Yet far too many people fail to grasp how science works as a discipline with its own philosophy and rules, and value it only as a source of useful gadgets. This is not their fault; it is the fault of the way science is taught. It is vital that we improve the way that science is taught because science offers the only means we have of tackling the looming world problems of climate change and the consequences of ever-rising numbers of people to house and feed.

I have emphasised in this book that science is not the coldly rational route to certain knowledge it is commonly thought to be, but an open-ended method of enquiry based on the assumption that the physical world is the only world there is. The science of biology teaches us that we are the evolutionary products of this natural world, and are thus subject to its unvarying regularities. We lose sight of this fact at our peril, because the roots of our thinking and behaviour lie in our evolutionary origins. We need to study those roots in order to rise above the limitations set by the undirected process of evolutionary change. There is very real danger that irrational thinking will threaten civilization and even human existence. We must resist what the late astronomer Carl Sagan called "the abject surrender to mysticism". Our future depends on how well we understand what we are, where we come from and the nature of the world we live in.

J. Ellis, *How Science Works: Evolution*, DOI 10.1007/978-90-481-3183-9,
© Springer Science+Business Media B.V. 2010

Further Reading

1. *The Demon-Haunted World: Science as a Candle in the Dark.* Carl Sagan. Published by Headline Book Publishing, 1996. ISBN 0 7472 5156 8. This book contains a personal plea for the need to use scientific thinking to safeguard democratic institutions and to combat the growth of irrationality in advanced societies.
2. *The Reason of Things: Living with Philosophy.* Anthony C. Grayling. Published by Phoenix, 2003. ISBN 978-0753817131. This book is a very accessible and thought-provoking discussion of the topics you need think about if you want to live a 'considered life'

Definitions

adaptation those properties of an organism that enable it to survive and reproduce in its natural environment.

agnostic a person who thinks that nothing can be known about the existence or non- existence of the supernatural.

anthropic principle the collective name for a group of ideas that assert that physical and chemical theories about the Universe must take into account the existence of human life.

archaea a group of prokaryotic organisms distinguishable from bacteria by several biochemical properties.

assertion the declaration that something is true.

assumption the position that something is true for the purpose of argument or action.

atheist a person who does not believe in the existence of the supernatural.

bacteria a group of prokaryotic organisms, distinguishable from archaea by several biochemical properties.

belief a statement of faith that an idea is true or important, whether or not there is testable evidence for it.

biodiversity the existence of many different types of organism.

biogeography the study of biodiversity across all regions of the Earth.

biosphere that part of the Earth that contains living organisms.

Cambrian explosion the relatively sudden appearance of muliticellular organisms about 540 million years ago in the fossil record.

crystallins transparent proteins found in eye lenses.

cyanobacteria a group of photosynthetic bacteria.

deism the belief in a supernatural agent who created the Universe but no longer interacts with it.

DNA deoxyribonucleic acid.

ecosystem a system of interacting organisms and their environment.

elongation factor a protein required for ribosomes to synthesize polypeptides.

empirical derived from observation or experiment and not from what someone tells you.

endosymbiosis one type of cell living inside another type of cell without harming it, and possibly providing some benefit.

enzyme   a biological molecule that catalyses a chemical reaction; most enzymes are proteins but some are made of RNA.

eugenics   the idea that humans should take steps to improve their genetic inheritance.

eukaryotes   organisms whose genome is surrounded by a nuclear membrane, thus separating transcription from translation.

evolution   the change in genetic composition of populations with time.

extant organism   an organism that occurs today.

extinct organism   a species that has completely died out.

eyespots   structures that detect the intensity and direction of light but are unable to form an image.

facts   in science, facts are observations that are empirical, repeatable, and shareable by everyone.

faith   belief in religious doctrines.

fitness   in evolutionary theory, fitness is defined as the mean number of offspring left by an individual, relative to the number of offspring left by an average member of the population.

genes   regions of DNA that encode RNA and protein molecules.

genetic code   the relationship between the sequence of bases in DNA in a gene and the sequence of amino acids in the encoded protein.

genetic drift   the change in gene frequency between generations caused by random sampling effects.

genetic system   any system that contains DNA, the enzymes to transcribe the DNA into RNA, and to translate the sequence information into proteins.

genome   the total genetic information in a given organism.

homology   in Darwin's time, "homology" described similar organs in different species, but today is often used to describe structures or molecules that are evolutionarily related.

hypothesis   an imaginary but testable speculation that might explain some facts.

intentionality   the tendency to interpret events in terms of purpose.

lateral gene transfer (LGT)   any process in which an organism incorporates genetic material from another organism, without being a direct descendent of that organism.

macroevolution   evolutionary change which produces new species.

messenger RNA   the product of transcription that is used by ribosomes to make proteins.

metabolism   the totality of chemical and physical reactions occurring inside organisms.

metaphysical naturalism   the assertion that the supernatural does not exist.

methodological naturalism   the assumption that scientists make that all that exists is the physical world that is characterised by unvarying regularities ("laws of nature") that can be studied by observation and experiment.

microevolution   evolutionary change occurring within species.

monotheism   the belief that there is only one supreme supernatural agent.

mutation   a change in the base sequence of DNA.

natural selection   changes in the genetic composition of a population due to differences in survival and reproduction.

naturalism   the assumption that everything there is belongs to the physical world we all aware of and which behaves according to unvarying regularities.

NOMA   an acronym for "non-overlapping magisteria", an idea suggested by Steven Gould that asserts that religion and science deal with different areas of human experience and thus cannot comment on each other's concerns.

Occam's razor   When several different explanations of a body of evidence are possible, prefer the one with the smallest number of assumptions, not because it is more likely to be correct but because it is the best way to proceed; the defining principle of science, also known as the law of parsimony.

occasional theism   the ability of some scientists to switch between naturalistic and supernatural types of explanation.

photosynthesis   the conversion of absorbed light energy to chemical energy.

plastid   a type of membrane-bound organelle found inside all plants, the most obvious and the most studied being the chloroplast.

polypeptide   a chain of amino acids in a defined sequence.

polysome   a molecule of RNA bound to more than one ribosome.

polytheism   belief in more than one supernatural agent.

prokaryotes   organisms whose genome is not surrounded by a nuclear membrane, enabling translation to be coupled with transcription.

protein   a molecule consisting of one or more polypeptide chains.

pseudogenes   genes that have lost their original function due to mutation.

random mutation   the observation that which particular mutations occur is unrelated to their effects on evolutionary fitness.

reason   the intellectual faculty by which conclusions are drawn from premises.

regulatory gene   a gene that encodes either a protein or RNA molecule that binds to another gene or genes and controls it or their expression.

religion   belief in a superhuman controlling power or powers, existing in an invisible supernatural realm and entitled to obedience and worship.

rhodopsin   a membrane-bound protein-retinal complex that absorbs light.

ribosomes   universal intracellular structures that synthesize proteins.

RNA   ribonucleic acid.

science   A set of ideas about the Universe based on empirical evidence, the use of Occam's razor, and the assumption that natural events have only natural causes.

secularism   the assertion that governmental institutions and policies should exist separately from religious beliefs and practices.

separate creationism   the hypothesis that species arose separately and independently by natural means: should not be confused with "creationism", "creation science" or "intelligent design", which are religious assertions.

species   a population of organisms that can potentially or actually interbreed; applies principally to eukaryotes.

supernaturalism   the assumption that beyond the obvious physical world there lies another invisible world containing one or more active agents.

theism   the belief that at least one supernatural agent created the Universe and continues to interact with it.

theory   in science, a theory is a coherent conceptual model that explains whole sets of facts and withstands predictions that could falsify it.

transcription   the enzyme-catalysed process by which one strand of the DNA of a gene is used as a template for the synthesis of a molecule of RNA with the same base sequence.

transcription factor   the product of a regulatory gene.

transformism   the hypothesis that all species arose independently but changed with time.

transitional fossils   fossils that show a combination of features from both their presumed ancestors and their presumed descendants.

translation   the process by which ribosomes use the base sequence in a molecule of messenger RNA to synthesize a polypeptide chain with a defined amino acid sequence.

uniformitarianism, principle of   the assumption that natural processes observed to be operating in the present also operated in the past.

unvarying regularities   another name for "laws of nature", used to avoid the misinterpretation that laws necessarily imply a lawgiver.

vestigial structures   structures that are reduced in complexity and function compared to similar structures in other organisms.

Wallace Line   an imaginary line in the ocean that indicates the separation of land species between the southeastern and northwestern parts of Indonesia.

Suggestions For Discussion Topics

1. What sort of empirical observation would persuade you that Darwin's theory of evolution is false?

2. Can you think of any ways of explaining the world other than naturalism and supernaturalism? Can you test any of these alternative explanations?

3. Discuss the four postulates that Darwin made in order for natural selection to operate (see Figure 3.6). What would be the effect on a population if postulates 1, 2 and 3 are correct, but postulate 4 is incorrect? Could natural selection operate if postulates 1, 3 and 4 are correct, but postulate 2 is not?

4. What is your view of those religious scientists who alternate between naturalistic and supernaturalistic explanations of the world? Does consistency of behaviour matter?

5. Discuss why mutation is random, but natural selection is nonrandom.

6. Conduct a survey of your classmates about what they believe about the nature of the world, and why they believe it. Then conduct another survey of what they think other people believe and why they believe it. Compare the surveys with each other.

7. Can you think of any empirical observation you could make or experiment you could conduct that would persuade you that supernatural agent(s) exist? Are you aware of any such observations or experiments?

8. Assume for the sake of argument that evolution is the invention of a supernatural agent, rather than a natural process. From your knowledge of how evolution works, what might you deduce about the character of that agent?

9. Compare the different lines of evidence for evolution outlined in Chapter 4. Which do you regard as the strongest evidence and which the weakest? Can you think of any other ways in which the theory of evolution could be tested?

10. Suppose that mutation is not random, but directed to create useful adaptations. From your knowledge of how the genetic code is used to make proteins, how might you explain such a hypothetical process of directed mutation, without invoking supernatural agents?

11. Some religious scientists argue that the 'laws of nature' have been created by a supernatural agent, while atheistic scientists point out that natural laws, by definition, have only natural causes. Which of these positions makes more sense to you? Explain why.

12. Suppose that no evidence has been found to support the idea of evolution. Given that both mutation and natural selection are observable facts, how might you explain this absence?